# REMOTE SENSING FOR LAND RESOURCE PLANNING

# REMOTE SENSING FOR LAND RESOURCE PLANNING

V.K. SHARMA

CONCEPT PUBLISHING COMPANY, NEW DELHI 110059

ISBN 81-7022-331-8

First Published 1991



*Published and Printed by*
Ashok Kumar Mittal
Concept Publishing Company
A/15-16, Commercial Block, Mohan Garden
NEW DELHI 110059 (India)

*Lasertypeset by*
Microtech Advance Printing Systems (Pvt.) Ltd.
H-13, Bali Nagar
NEW DELHI 110015

# PREFACE

JUDICIOUS USE of land resources is most essential for realising their productivity potential, for maintaining a satisfactory socio-economic growth rate overtime, and for minimising environmental degradation due to developmental activities. These development processes are essentially 'environment-sensitive' input parameters required for developing the land resource potential. Their irrational application becomes a risk and counter-productive to the quality of environment, which eventually leads to diminishing returns from the limited land resources. It is necessary, therefore, that development plans are perceived and implemented on scientific principles of land evaluation.

It is common knowledge that all the similar-looking parcels of land do not hold common promise for sustaining a given economic activity. Further, the ever-increasing diverse needs of growing population come in direct conflict with the suitability and optimal productivity of land resources. This requires classification and information on the geomorphic character and adaptability for dialectic adoption of land for uses to which it is best suited. A number of land classifications from specific to general purpose uses of terrain have, therefore, been proposed from time to time, and documented in literature. The principles, merits and limitations of some such schemes have been briefly summarised in the present work. It may here be mentioned that a comprehensive classification of land resources for all possible types of land-uses has not yet been achieved. In the present study, land resources in parts of Sonipat, Rohtak, Bhiwani, Mahendragarh and Gurgaon (including Faridabad) districts in southern Haryana, and a fraction of the Union territory of Delhi have been classified into disparate tracts of fluvial, aeolian and inland lake evaporite environments dispersed over four spatially non-contiguous tracts covering 5483.34 sq km area. This classification is based on black and white aerial photographs on an approximate scale of 1: 25,000. The interpretation is established on standard photo-elements and supplemented with limited ground truth control and synthesis of the data. Each of the above geomorphic environments have been subsumed into parcels of land homogeneous in respect of form, material composition and related geomorphic processes. These environmental resource units are geobiological complexes that are characteristic of typical abiotic and biotic associations. Further, they also possess similar resource potential and pose identical problems for a given use of the land resource. In this methodological framework, each environmental resource unit derives its intrinsic properties from functional interactions of climate, process, geology, geomorphology, and hydrology as abiotic elements. These lateral and vertical interactions set specific limits to the potential and productive levels of the resource units which in turn govern the processes of land resource development through the application of input parameters of relevance to the development activity.

Documentation of land resources and their inherent characteristics form the basis for planning the resource-base. Planning is a deterministic action with a futuristic view. It can either be short or long-term proposition which translates the basic theoretical knowledge into a practical reality matching that of the real-world situation. In the context of predominantly agrarian infrastructure, the planning is to be oriented for improving the agricultural production and maintaining its satisfactory growth rate from the given land that has been tilled for centuries, or has been brought under agricultural use only recently. There is no denying the fact that the desired socio-economic development in India can be brought about only through agricultural sector which supports the livelihood of some 80 per cent of the population. Socio-economic development and rural development are, therefore, synonymous when such a large population base is involved in the planning process.

Alternative land-use planning suggested in this research is based on the integration of geoenvironment potential of the resource units with the land-use data. Whereas the land cover data have been derived as a corollary to the classification of geomorphic environments, the agricultural land-use data were obtained from the unpublished revenue records available at tahsil headquarters of the respective regions. The relationship between the properties of environmental resource units, land cover, and agricultural land-use coupled with other infrastructural facilities for the support of agriculture suggests whether the existing cropping pattern and crop yields are in harmony with the geomorphic environment. Alternatively, suggestions have also been made to replace the given crop with a crop or crops most suited to the environment. Ways and means to augment crop yields, and the processes leading to the degradation of environment have been highlighted.

It is hoped that this research based on remotely-sensed data on land resources would be useful for planning areas with similar geological environments. The methodological framework discussed herein applies not only to agricultural land-use planning but also to the resource development in which other priorities for use of land may be more important.

V.K. SHARMA

# ACKNOWLEDGEMENTS

I WISH to express my gratitudes to the University Grants Commission, New Delhi for the award of this research. The completion of this research project would have been difficult, indeed, without the necessary facilities provided to me by Professor Jasbir Singh, my colleague in the department of Geography. Thanks are also due to Dr. G.P. Bhargava of the Central Soil Salinity Research Institute, Karnal (Haryana) for discussion on many aspects of this research. This is to acknowledge also the assistance of Mr. Bhupender Singh Kundu, Junior Research Fellow, in interpreting a few aerial photographs. Finally, I wish to thank Dr. R.L. Sharma for cartographic advice, and Mr. Achla N. Sharma for typing the manuscript.

V.K. SHARMA

# CONTENTS

# LIST OF FIGURES

# LIST OF TABLES

# 1

# INTRODUCTION

NEARLY THREE-FOURTHS of the population in India lives in villages, and is directly or indirectly dependent upon agricultural pursuits for livelihood. The agricultural sector is, however, not able to sustain the large rural population due primarily to generally low-levels of agricultural productivity, fragmentation of land-holdings on account of social and cultural traditions, and inadequacy of infrastructural facilities for the support of agricultural pursuits. Growing levels of the higher education among rural youth have also lured the locally available potential agricultural labour force to white-collared jobs in towns and cities, thereby creating a void in the labour requirements of agricultural sector in the still labour-intensive agriculture that is so well characteristic of much of the subcontinent. This deficiency is more or less compensated by the migration of agricultural labour force from regions of lower agricultural productivity to the islands of agricultural prosperity, such as the States of Punjab and Haryana, and the western parts of Uttar Pradesh.

The development plans are aimed at achieving a certain economic growth rate over a period of time. The planners rely mainly on the socio-economic parameters for formulating planning strategies and their subsequent implementation by the decision-makers at the political level. In so doing, the other parameters of natural resources that are of immense significance to any development plans are relegated to a secondary importance, or generally not considered significant to the planning process. Consequently, optimum and judicious utilisation of these resources does not form an initial component of developmental plans. Incidentally, it is taken for granted that any socio-economic growth rate achieved through the planning process is due to effective utilisation of the natural resource base. In such a circular argument, the long-term effects and risk-factors of the envisaged plans do not normally conform to the real-world situations.

Planning in the Indian context, or for that matter in any other similar agrarian economy, can serve its intended purpose if it is oriented for agricultural land-use planning. The pattern of agricultural land-use is, to name a few, developed from the pattern of operational size of land holdings, the agricultural density, the food habits, and the infrastructure for the support of agricultural operations. The planning process strives to achieve sustained socio-economic growth rate through raising the levels of agricultural productivity from the land that has been tilled for centuries, or has recently been brought under agriculturally-related infrastructural influences. The application of inputs, such as irrigation, mechanisation, pesticides and insecticides, fertilisers, and high yielding varieties, is intended to exploit the potential of land for agricultural purposes. In consequence thereof, the carrying capacity of land in relatively underdeveloped and marginal lands, such as in southern Haryana and elsewhere in dryland environments, has increased manyfold. This planning perspective wilfully ignores the long-term adverse effects of development on degradation of environment. It is not surprising, therefore, that the emphasis on planning now-a-days has shifted from mere exploitation to judicious exploitation and development of the environment in its total character.

Land resources on earth, the only suitable habitat, are finite. The manner in which land is used largely determines the physical, social, and economic environment of the population. The ever-increasing pressure of population, irrational exploitation of resource base for quick, short-term gains and economic prosperity, and intensification of agriculture with inappropriate technology — especially in the developing world — have caused deterioration of environment with cumulative effects. The most crucial question facing the planners is, therefore, the integration of the process of agricultural development with the protection of environment, maintaining the character of land, and retaining the cultural traits of the area, while at the same time satisfying the essential requirements of the present population and that of the generations to follow.

Land-use planning is primarily concerned with the spatial pattern of land-uses and their impact on landscape from a utilitarian viewpoint. Land-use planning is, to quote Spangle and Associates *et al.* (1976), an integral component of the comprehensive planning that treats the future development of an area in terms of all major determinants of growth and change.

The land-use planning process consists of five conceptually distinct and continuously interacting phases. To quote Spangle *et al.* (1976), these are:

(1) identification of problems and definition of aims and objectives;
(2) data collection and interpretation;
(3) plan formulation;
(4) review and adoption of plans; and
(5) implementation of plan.

The aims and objectives of land-use planning at

local, regional or national level should conform to the national policies on resource conservation and development. Policy objectives related to preservation of environment are unfortunately not closely adhered to in the developing countries, where the shear magnitude of the size of poverty has caused extensive environmental degradation. Further environmental degradation is also on the anvil due to increase of human population and that of cattle. The urgency of socio-economic uplift has often created a situation where the process of development at the cost of conservation of environment has been allowed to proceed unabatedly. It is high time now for the policy-makers to seriously think that the cost of environmental degradation is far more in excess of the long-term consequences of such ill defined, conceived, projected and implemented socio-economic development plans based on utilisation of land resources.

Wise use of land and its conservation require diverse data on geology, climate, topography, pedology, hydrology, and other large number of related disciplines for interpreting the natural materials, features and processes of land, and risks and geological hazards to be encountered during the formulation and after the implementation of plan. Data on these parameters is an essential prerequisite for classification and evaluation of terrain and other related physical components of the environment that set limits for specific and general purpose uses of land resources.

Formulation of a land-use plan is usually based on assessment of land capability of the area under consideration. Planners are agreed that land capability portrays the natural opportunities and constraints in the use of land. Further, appropriate weightage is also given to the proposed socio-economic growth rate within a given political set-up. Proposals, that involve resource conservation and development, are suggested for suitable and desirable changes in land-uses that are subsequently subject to assessment of their impact on environment, economic and social change. It is also desired that hazardous areas be identified and evaluated for level of risk. Further, alternative measures for mitigating hazards are also evaluated for both environmental and economic impacts. The envisaged plan is flexible enough to incorporate revisions, if any, in land-use planning. In the context of a land-use plan for rural India, the proposed changes, as referred to above, should not only enhance the quality of environment but also arrest, and reverse the flow of rural population to urban areas where slums develop as inseparable component of urban landscape. The adverse effects of squatter dwellings on the environment, and social implications of their mushroom growth are, however, beyond the purview of this research.

The plan is reviewed and adopted after official approval to it has been sanctioned by the various decision-making bodies of the government.

The plan is implemented in phases by acquiring land, and by conducting the neccssary studies on terrain and related physical aspects by trained personnel. The plan progress is constantly monitored, and reviewed at the time of completion of each phase of the plan. Changes, if required, are incorporated in the subsequent phase of the plan. Since there are no holidays in the planning process, the planning continues afresh for the same area after the completion of the contemplated plan period.

Preparation of a land-use plan is a complex problem and requires a team approach. In this context, the earth scientists have a special role to play in that they provide valuable data on inventory and prospects of land resources for developmental purposes. At the outset, it may be emphasised that all the land is not the same in terms of its morphologic and morphometric attributes, and in the physical and chemical properties of earth materials. Therefore, contiguous and non-contiguous parcels of land possess specific limitations in utilisation for various purposes, and each to be productive requires specifically formulated conservation practices. Depending upon the specific terms of reference of the proposed plan, earth scientist has the responsibility of collecting available earth science information (topography, geology, soils, and hydrology), acquiring necessary new data, and interpreting these data for land capability classification. To quote Spangle *et al.* (1976), the last named spatial pattern should ideally correspond with the natural capability of a land unit (parcel of land homogeneous in respect of its physical environment) for specific potential use. The earth scientist has the additional duty of preparing a relevant technical report, interact with specialists in other disciplines related with the planning process (economics, demography, sociology, agronomy, political science, and engineering, for example) during the phases of land-use planning, and to suggest a prognostic approach for mitigation of geological hazards.

Resource management requires understanding of environment, rational analysis, and decisive action (Gregory, 1973). Environmental data comprise of qualitative observations and quantitative measurements. These data can be acquired from many sources, including remote sensing. Satellite imagery and aerial photographs provide a uniform data-base for rapid assessment and interpretation of the physical attributes, distribution, utilisation and productivity of natural resources, and in the analysis of the interaction between earth resources and land-use. Remote sensing also provides a cost-effective method of monitoring the environmental dynamism.

Remotely-sensed primary geobiological data are based on personal decisions and judgement. Hence, they may not always be unequivocal as the aquired data emphasise only selected attributes at the expense of others. The accuracy of classification, and the inferences derived from the study, depend on relative scale of observation, context, and interpretive skill and subject expertise of the photo analyser. Terrain is the most imposing aspect of the physical environment that lends itself to analysis by remote sensing techniques. The spatial data on terrain may be analysed by adopting empirical-quantitative and explanatory-descriptive approaches. Aerial photographs that place an infinite access to the wealth of detail about the regional character, components of terrain and processes, have been extensively used worldover for terrain analysis. Further, the interactions between terrain properties, type of soils, hydrologic attributes, and biotic associations can be discerned with greater precision.

Landforms, the most fundamental of terrain attributes and diagnostic of geomorphic environment, provide basic source of information for terrain analysis and its application to regional development and planning. Glennie (1970), Way (1973), and Sharma (1988) have, for example, used aerial photographs for the identification and interpretation of landforms that provide fundamental data for land planning, landscape architecture and similar projects. Landforms can be subsumed into the smallest, relatively homogeneous elements of geomorphic landscape through the progressively higher hierarchical orders of landform groupings. In scale of investigation, these classifications cover areas of less than one sq km to large geographical regions, and finally to assemblage of landforms on a global perspective. The methodological framework of terrain classification has been discussed in detail at a later stage in this chapter.

## Study Area

Dispersed in four segments, the study region in southern Haryana and a part of the Union territory of Delhi covers some 5,438 sq km area (Fig. 1.1). In Haryana, these areas cover parts of Sonipat, Rohtak, Bhiwani, Mahendragarh, Gurgaon, and Faridabad districts; the last named district being represented in the undivided district of Gurgaon.

The study area was selected on the bases of the availability of reasonably good quality black and white panchromatic aerial photographs on an approximate scale of 1: 25,000, and Survey of India topographical maps on 1: 50,000 scale for the corresponding runs of aerial photographs. Representing distinct geomorphic environment(s), these four areas are listed in Table 1.1 with reference to the Survey of India topographical maps, year of photography, and area dimension.

## Methodology

Inventory of land resources is an essential component of resource management in less-developed areas, and in those areas that have haphazardly developed of their own over a long period of time. The data on land resources provide basic information for land management and planning decisions that may be undertaken at several hierarchical levels of complexity (Ollier, 1977). A land evaluation map is useful to agronomists, hydrologists, engineers, resource managers, and others for analysing multifaceted practical problems arising out of the use of land.

TABLE 1.1: Area Reference

| Sl. No. | Region | Survey of India topographical maps | Year of aerial photography | Geomorphic environment | Surveyed area (sq km) |
|---|---|---|---|---|---|
| 1. | Sonipat area | $53^C/16$, $53^D/13$, $53^G/4$ and $53^H/1$ | 1969 and 1979 | Fluvial environment | 1507.16 |
| 2. | Rewari area | $53^D/7$, $53^D/8$, $53^D/11$, $53^D/12$ and $53^D/15$ | 1965 and 1979 | Dryland environment | 2887.75 |
| 3. | Nuh area | $53^D/16$ and $53^H/4$ | 1979 | Dryland, and Inland lake evaporite environments | 592.87 |
| 4. | Ferozepur Jhirka area | $54^A/13$, $54^E/1$ and $54^E/5$ | 1976-77 | Fluvial, Dryland, and Inland lake evaporite environments | 495.56 |
| Total area (1-4) | | | | | 5483.34 |

Land resource map represents a synthesis of the geomorphic landscape and of the ecological complex. The evaluation of landscape intangibles for land-use planning is, as noted previously, a complex exercise concerning economics, social consequences, benefits, and hazards following a given use of the land (United Nations, FAO, 1976). It is, therefore, necessary to evolve a sound methodology to ensure that land resources are evaluated, used and conserved in a manner consistent with the needs and aspirations of the society.

A perfect, all comprehensive system of terrain classification is yet to be achieved. Consequently, many types of land classification have been proposed for a variety of purposes, and applied with fair amount of success in the evaluation of land performance for specific and general purpose uses of land. The potential usefulness of terrain classification stems from the fact that many kinds of land-uses are constrained by the same quality of land. Even world's ecosystems are shown by

Budowski and Miller (1973) to be intimately related with the physical properties of terrain.

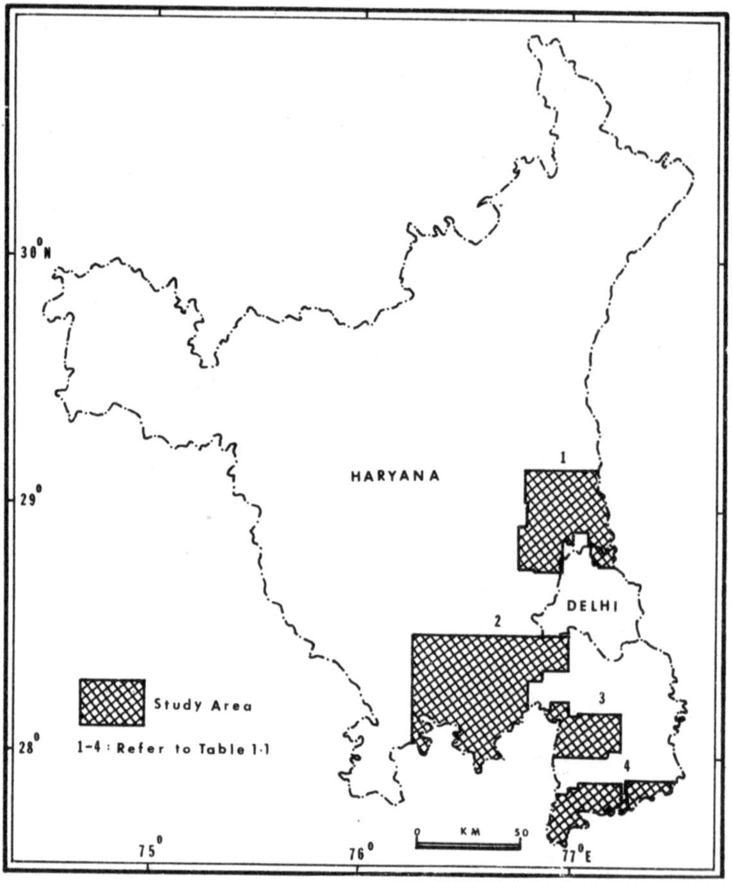


FIG. 1.1: Study Area.

Terrain classifications are based on qualitative observations and quantitative measurements. In qualitative method, terrain may be subsumed into types based on morphologic attributes of forms. On the other hand, quantitatively defined observations are objective and comprehensive, defining variables of terrain on identifiable facts that embody interpretation of their meaning as well. These parametric values are amenable to statistical analyses, and for examining the interrelations that exist among the various components of terrain. It is no denying, however, that these observations are subjective as they are specific only to the context of study. In this context, the classification proposed by Van Lopik and Kolb (1959) is particularly a good illustration of how parametric terrain variables can be grouped to differentiate terrain types in a desert environment. In this classification, the typical variables are subsumed into five characteristic types as:

1. Aggregate or general terrain factors – physiography, hypsometry, and landform-surface conditions,
2. Geometry or form factors – characteristic slopes, relief and plan profile,
3. Ground factors – surface rock, soil type and soil consistency,
4. Surface roughness and micro-relief – quantitatively described and analysed, and
5. Vegetation – type, distribution and density. With parametric data, it is usually possible to develop regionalisation of landforms in terms of their attributes and spatial associations.

The conceptual framework of classification systems has been discussed in the United Nations, FAO (1974) Report. In this context, approaches to land evaluation may be subsumed into two basic types as terrain-based and soil-based classifications.

Wright (1972) has summarised the various methodological frameworks for terrain-based land classifications. These can be broadly grouped into two basic types that concern with: (i) The total description of landform attributes, and (ii) differentiation of terrain on some selected geomorphological properties or combination of properties considered relevant to a specific use of land. In the former explanatory descriptive approach, landforms are mapped with regard to morphology, morphometry, morphogeny and morphochronology. The information presented on a single map with the help of symbols and/or colour scheme offers more confusion in the interpretation of map, and the significance of mapped landform attributes to land utilisation. In this approach, dynamics of land evaluation and not the quantitative information on form and material composition forms the focal theme which obliterates the usefulness of geomorphic maps to the evaluation of land performance *vis-a-vis* land utilisation.

Differences in emphasis from genetic-chronological types of geomorphic maps as in France and Poland, to empirical-quantitative morphologic types such as in United Kingdom, to a blend of both types in Belgian scheme clearly reflects the serious concern of geomorphologists in evolving a suitable and a useful method of terrain classification that can have wide ranging applications in land resource utilisation and management. The morphological maps describe the form of surface morphology in plan form by the facets of land surface. The facets are defined by changes and breaks of slope (Bridges and Doornkamp, 1963). Additional information, such as on slope aspect, depth to water-table, nature of slopes – whether erosional or depositional – may also be added (Waters, 1958) to strengthen the significance of the approach to land evaluation and utilisation. The extent of detail mapped requires intensive field-work that obscures the very utility of the maps for the intended purpose. In recent years, applications of remote sensing techniques have, however, greatly reduced the effort and time in arriving at such maps.

The land system approach developed by the CSIRO Division of Land Research, Australia treats the land as an 'organic whole' or an 'ecological complex' throughout which, to quote Christian and Stewart (1952), a recurring pattern of topography, soils and vegetation can be recognised. Conceptually, a land system is defined by Christian (1958) as a group of closely related topographic units (land units), usually small in number, that bear a common genesis. The properties and parameters of the various land units comprising a land system are, however, not examined in detail. A land unit is recognised as a particular landform which in its occurrences is associated with the same group of soil and vegetation communities (Christian, 1959). The terms topographic units, land units and landforms are used interchangeably which does not clarify whether a unit is a segment of slope or a landform *per se*. Further, the physical attributes of land units are only casually examined. It may be noted that the main emphasis in this approach is to identify and delineate boundaries of land systems which may be termed as major divisions of the earth's surface of given geological environments. The land system approach however provides a useful framework within which the range of potential and possibilities for sustaining a given use of land can be assessed.

Land system approach using remote sensing techniques is being increasingly applied to land resources evaluation and planning in India as is evident from a large number of papers presented in different symposia (see, for example, *International Symposium on Arid Zone Research and Development*, Jodhpur, February 14-18, 1978), and many other research publications. Raghavswamy and Vaidyanadhan (1980), for example, have assessed capability and limitations of land systems in Visakhapatnam district of Andhra Pradesh. Further, current agricultural land-use has been described in order to recommend suitable changes in agricultural land-use pattern in relation to terrain properties of individual land systems. Some other studies, such as by Raghavswamy (1982) and Ramaswamy *et al.* (1984), also attempt inventory of land resources using the land system approach.

In contrast to the total description of terrain, much effort has also been directed to the evaluation of terrain for specific use. This approach differs from the previous parametric attempts in a way that physical limitations, degree of freedom and preferences in the use of terrain for specific purpose are evaluated by considering either one dominant criterion or a group of interrelated aspects relevant to the use for which terrain properties are being analysed. The Military Engineering Experimental Establishment (MEXE), now known as Military Vehicles Experimental Establishment (MVEE) at Oxford, U.K., has developed a scheme of terrain classification for some specific purpose; the thrust being to analyse the suitability of a site for military purposes. In this context, terrain is defined in terms of morphology, surfacial material and water regime, and spatial variations in the three intrinsic properties. Facet, the basic unit, is the largest tract of terrain that can be mapped on 1: 50,000 scale, and is homogeneous in terms of the above-referred attributes. Such a unit of landscape is, therefore, suitable for moderately extensive land-use or constructions. Subfacets constitute those parts of facets which are anomalous with regard to morphology, material composition and water regime. Regularly occurring groups of facets constitute 'recurring landscape patterns' (RLPs). The facets occurring in one RLP always exhibit the same spatial interrelations among themselves. Several MEXE (MVEE) reports on the subject are available in mimeo forms. Beckett *et al.* (1972) have discussed the methodology and implications of the classification scheme for resource appraisal and utilisation of land resources, with examples from desert and temperate lands. The system seems to work well in stable ecosystems, but it is difficult to apply in developing countries and in tropical areas where land-use has been consistently evolving in the recent decades (Millard, 1972).

The MEXE/MVEE studies of terrain analysis are designed to answer two questions, viz. (i) How homogeneous are terrain facets in one land pattern? (ii) How dissimilar they are in their occurrences in other land patterns? Despite the basic argument that facets are genetically related to common process does not always hold (Beckett *et al.*, 1972). If the study is to serve its intended purpose, time consuming field and laboratory investigations are required to analyse the physical properties of the facets.

Impact of some specific terrain characteristics on land-use may also be investigated from the viewpoint of limitations imposed on the use of land by the spatial pattern of geomorphic hazards. Suitability of terrain for land-use is, thus, evaluated in terms of the nature, frequency and magnitude of geological events. In one study, Dow *et al.* (1981) employed mass-movement phenomena to classify the suitability of mountainous terrain for various land-uses. Such an approach provides a useful framework for the analysis of land-use oriented problems, and for land-use planning. The importance of geomorphic hazards associated with specific landforms is being increasingly appreciated (see, for example Cooke, Goudie and Doornkamp, 1978). In the light of above-referred studies, engineering decisions, and decisions regarding land-use planning can now be made more judiciously.

The methodological framework of soil-based land

classifications has been summarised in the United Nations, Food and Agricultural Organisation Report of 1974. Accordingly, parcels of land are grouped in terms of quantitatively evaluated soil potentiality, productivity or suitability status. The land capability classification of the United States Department of Agriculture, for example, provides for grouping of soil units that respond in a like-manner to management, limitations, risks, and adverse effects. Economic land approach is modified land capability classification that perceives of both biotic and abiotic conditions controlling yield of crops. The productivity classification is largely based on soil fertility status of ecologically homogeneous parcels of land. This ecological approach evaluates all those inherent parameters of land, soil and water which together govern the productivity of a given piece of land. Land suitability classification refers to the suitability of soil for a given use of land. The scope of such classifications, for example, can also be extended to devise schemes suggesting suitability of irrigation for a given soil.

The Command Area Development Authority, Haryana has begun investigating the problems of development within the canal command areas. Besides analysing the agricultural economics and other socio-economic parameters of development, the organisation has also prepared maps showing soils, soil irrigability, land capability and groundwater quality. The dynamics of rural development as envisaged by CAD is, however, biased in favour of the socio-economic variables.

In this research, land resources have been evaluated on geomorphic criteria involving geomorphic environment and the three-dimensional morphologic attributes of landforms. Terrain classification based on geomorphic criteria is, to quote Ollier (1977), 'firm in ground truth and adaptable to many purposes'. Such an approach is also supported by Glennie (1970) who classified the terrain in the deserts of Arabian Gulf into areas of marine, aeolian, fluvial, and evaporation environments and their associated landforms. Recently, Rawat *et al.* (1988) and Sharma (1988) have examined the usefulness of analysing geomorphic environment for exploration and utilisation of land resource potential for development. Sharma (1988) classified the fluvial terrain in parts of Haryana and Delhi into related landforms. In such a scheme of classification, the subenvironments within a given environment may be identified from the morphology, material composition, and associations among forms and their attributes. This approach has the additional advantage in that the aspects of surface morphology, geology, soils, vegetation, and hydrology become functionally and genetically related with each other, and with the environment of erosion, transportation, and deposition. The units of land surface with homogeneous characteristics in respect of the above-mentioned aspects and, consequently, in geobiological environment have been referred to as environmental resource units. These environmental resource units possess similar conditions of sensitivity and stability that can be evaluated for general and specific uses of land.

The scheme of terrain classification presented in this research is, with few exceptions, similar to the land system approach of the Land Division of the Council for Scientific and Industrial Research Organisation, Australia. The two schemes are compatible in respect of a genetic classification of terrain, and in the treatment of classified environment as an ecological complex. The land system approach, as mentioned elsewhere in this research, falls short of further analysis of land units or geomorphic attributes of the subsystems within a classified environment. The scheme of terrain evaluation discussed in this research, however, fully accounts for the subenvironments and their abiotic and biotic attributes. These geobiological resource units have been further analysed and classified into smaller environmental resource units for the interpretation of geomorphic hazards in relation to the natural and man-made fabric of the landscape. The sensitivity and stability of each resource unit can, thus, be integrated with the administrative process of land resource planning. It may here be mentioned that all such units are sufficiently large in size for a moderately extensive land-use.

Inventory of land and land-based resources also requires information on land-use and land cover. The data on these two aspects and thematic maps prepared therefrom are useful from the viewpoint of appraisal of economic and environmental problems facing the world today. Anderson and Witmer (1984) note that though land-use and land cover data do not provide complete information necessary for land resource planning and management, yet a standardised information on these two provide an essential element in any system for land resource planning and management. Land-use and land cover maps when compared with soil maps, for instance, provide a better understanding of the circumstances under which transformation of a given use of land to another type should or should not be permitted in the long-term interests of the society.

Using geomorphic criteria, conventional black and white aerial photographs were analysed by cerebral processing method for identification and delineation of various geological environments and subenvironments within, and associated landforms in the project area. These physical aspects of the environments have been recognised on shape, size, tone, texture, pattern, association, and convergence of evidence observed on stereopairs. Similarly, land covers – some of which reflect the potential and limitation of resource units – were also delineated and classified into units of abiotic and biotic landscape. All these boundaries were photographically reduced to superimpose on the corresponding Survey of India topographical maps. In conjunction with the data on land-use, these base maps have been interpreted for land resource evaluation. In this context only Level A land-use data from revenue records, and other publications have been used.

The boundaries of environmental resource units were field-checked along selected traverses. In case of Sonipat region, soil samples from root-depth were also collected, and analysed for selected physio-chemical properties of soils. For the remaining three areas, these data were obtained from several reports and publications, and integrated with the various aspects of the study. Other available collateral data, such as on lithologs, hydrology, groundwater and land capability surveys have also been extensively used for suggesting alternative land-uses.

# 2

# LAND RESOURCES AND LAND RESOURCE PLANNING IN SONIPAT REGION

## Introduction

SUSTAINED PRIMARILY by labour-intensive agricultural growth, the fluviatile Indo-Gangetic Plain has for centuries supported one of the highest density of population in the world. In the past, numerous diversified themes on man-environment interactions have been published for the Plain by States and smaller enumeration units, and for selected watersheds. The topographic descriptions of the Plain have usually been sketchy, and added more to empirical descriptive classification of terrain than to have projected its geomorphic evolution. The applications of such studies to an understanding of man-environment relationship are, needless to say, based on inadequate evaluation of terrain characteristics of the productive lands. The geological studies on the evolution of the Indo-Gangetic Plain, or a part thereof, have tended to be macro-scaled, but could be extended for a comprehensive interpretation of the various aspects of the terrestrial environment.

Covering some 1507 square kilometres, the study area in parts of Sonipat and Rohtak districts of

Haryana, and a part of the Union territory of Delhi extends approximately between 28° 46' to 29°10' north latitudes and 76°45' to 77°14' east longitudes (Fig. 2.1). The elevation in the Plain varies from 221 m in the north to 210 m in the south along the Jamuna river which forms the eastern boundary of the region. The elevation rises to 222 m in the southwest and to 228 m in the west. Within this framework, minor random elevation varions are, however, as prominent as would be typical of a fluvial environment of erosion and deposition.

The region experiences a mean annual rainfall of 60 cm which has a strong component of April through September summer season rainfall of 55 cm. The variability of rainfall is, to quote Rani (1978), given at 30 per cent for the mean annual, 40 per cent for the summer season, and 70 per cent for the winter period rainfall. With an annual potential evapotranspiration of 142-145 cm, the study area is characteristic of a semiarid environment, and perpetual drought throughout the year (Sharma, 1974). Supported by canal and minor irrigation networks, the extensive cultivation drives the economy of the region, and is the main stay of the livelihood of some 80 per cent of population directly or indirectly engaged in agricultural pursuits.

## Topography and Geology

The Indo-Gangetic Plain is a major geological basin, and a topographic division of India. The alluvium was deposited in a tectonic platform depression between the peninsular and extra-peninsular regions of India during the Quaternary period. Therefore, it is geologically young, and so is the topographic expression developed on the alluvial terrain. In geological literature, this depression of a regional magnitude is known as a 'foredeep' a 'syneclise', or a 'crustal structure of a negative character'. The thickness and variation of alluvial cover is said to be determined by the subsurface extension of tectonic lineaments of the peninsular rocks into the Indo-Gangetic plain. These buried basement features are, to some extent, reflected in the subsurface configuration and geohydrologic characteristics of the overlying alluvial terrain.

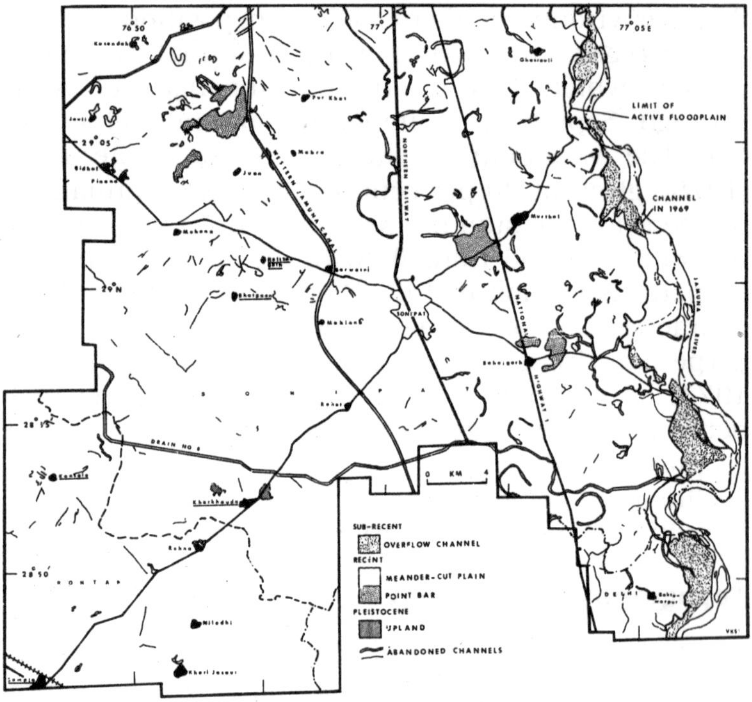


FIG. 2.1: Classification of fluvial environment and associated landforms.

Geddes (1960) construed that the Indo-Gangetic Plain is composed of several huge alluvial cones built by the major Himalayan rivers draining into the Bay of Bengal and the Arabian Sea. Further, the surface of the cone in northwest India, as elsewhere, was incised in the Holocene by the present-day streams. The part of the alluvium that has been extensively reworked by fluvial activity is referred to as the newer alluvium or the *khaddar*. The topography of this surface is expressed by numerous buried channels, palaeochannels, oxbow lakes, meander scrolls or point bars and meander-cut plain. The *khaddar* is reportedly rich in detrial materials like garnet and mica, and heavier lag deposits in abandoned channels. In fact, these features represent the areas of present, active floodplain, and the past floodplain of the river(s) traversing through the pre-existing alluvial tract. The *bangar* or the older Pleistocene alluvium occupies the higher topographic positions on the Indo-Gangetic alluvium not affected by the fluvial activity so well characteristic of the *khaddar*. Therefore, the *bangar* tract makes the highest terrace in this fluvial environment (Sharma, 1986). These two fundamental topographic divisions of the alluvial plain as 'newer' or Recent and 'older' or Pleistocene have been accepted by earth scientists (see, for example, Baweja and Sharma, 1978, and Dhankar *et al.*, 1983). The distributional pattern of 'newer' and 'older' alluvium recognised by Baweja and Sharma (1978) on 1: 1 million Landsat imagery does not, however, even roughly correspond with the one defined in this research on aerial photographs. Further, the limits of these two units of alluvium given by Dhankar *et al.* (1983) for Haryana, and their divisions, presumably based on topographic maps, into empirical terrain types also stand in contradiction to the one described in this research. Notwithstanding the discrepancies in the recognition of the Recent and Pleistocene alluvial tracts, these two basic divisions of the alluvium not only reflect the geomorphic events or processes, but also the characteristic topographies associated with each types of the alluvium.

Apart from the site-specific boundaries of the two topographic surfaces, one another basic difference in the chemical composition of the alluvium may also be cited. The 'newer' alluvial tract is, to quote Pascoe (1950), altogether free of salt accumulations, while the 'older' surface invariably contains typical deposits of calcium carbonate nodules or calchie at and extending some distance beneath the surface of an enormous thickness of alluvium. These concretions and salt-encrustrations are evaporites formed by the evaporation of soil-capillary and stagnant waters in the high thermal efficiency regime of the area. Locally referred to as *reh* or *kallar*, these salts are rich in carbonates, sulphates and chlorides of sodium, calcium and magnesium. However, a classification of alluvium into *khaddar* and *bangar* on the basis of salt efflorescence is not a valid criterion since, as discussed later in this chapter, large tracts of *khaddar* land have also been infested with calchie. Therefore, the geomorphic observations of reworked alluvium is the only suitable criterion for distinguishing the two surfaces which otherwise differ only slightly in other details. The sediment data from the Union territory of Delhi, however, suggest that the *khaddar* of the Jamuna overlies the sands of aeolian environment (Srivastava *et al.*, 1976). Therefore, the river deposited sediments post-date the aeolian sands in this area.

According to the tectonic map of India published by the Oil and Natural Gas Commission in 1968, the *khaddar* in the study area is restricted to between Jamuna and the northern railway line, while the area to the west of railway tract passing through the Sonipat town is shown to be of the older alluvium (see Fig. 2.1).

On the basis of geophysical and other data, the geologists have inferred that a large basement high of peninsular rocks is buried beneath a variable, yet of enormous thickness of alluvium in the area between the Jamuna and Ganga systems, and the rivers of Punjab draining into the Indus River (Rao, 1973). The thickness of alluvial cover over the basement rocks is not precisely known, but it may be anywhere up to 2 km thick at places. On the bases of subsurface stratigraphy and geological evolution, Sastri *et al.* (1971) suggest that two subsurface faults traverses through the Sonipat district in a roughly northeast-southwest direction. Further, a subsurface geological structure referred to as 'Delhi-Hardwar Ridge' has also been identified in the eastern part of the Sonipat district. This ridge in the district is inconspicuous as it is reportedly shallow in character and overlained by an enormous thickness of alluvial cover. The geomorphic implications of such geological structures in the evolution of drainage and topography are, however, not understood in clear terms at the present time.

**Discussion**

The Quaternary fluvial environment in the Sonipat region has been classified into three geological periods as Sub-Recent, Recent and Pleistocene, and the respective landforms of each period have been identified. These landforms are further classified according to their subenvironments into four major landforms (Fig. 2.1). Each of these landforms constitute a land resource unit or an environmental resource unit which in its various occurrences is homogeneous in respect of process, morphology and material composition and, consequently, in geobiological environment. Aerial photographs have been further analysed for land cover types (Fig. 2.2), and examined in respect of selected physico-chemical, properties of soils. A casual relationship between the Quaternary landforms, land cover and agricultural land-use data are developed to explain the potential and limitation of the above resource units to sustain the given use of land and to suggest an appropriate land-use under the prevailing environmental conditions.

**Fluvial Environment**

As shown in Fig. 2.1 and detailed in Table 2.1 the fluvial environment of erosion and deposition comprises of Sub-Recent overflow channels, Recent meander-cut plain and point bar, and Pleistocene upland tract. The distributional pattern of the features suggests major hydrographic changes possibly related to two river systems; the present Jamuna in the east, and the Sarasvati of the historic past in the northwest and west. It is also possible that Jamuna that may have drained into the Arabian sea through the Sarasvati-Ghaggar system had in the Recent times changed its course to become a tributary of the Ganga system, Numerous abandoned short channel segments dispersed throughout the study area attest to the presence of widespread reworked alluvium in place of the pre-existing older Pleistocene alluvial tract. It may, therefore, be inferred that the fluvial activity since the Holocene has not been restricted to areas adjacent only to the present-day stream courses.

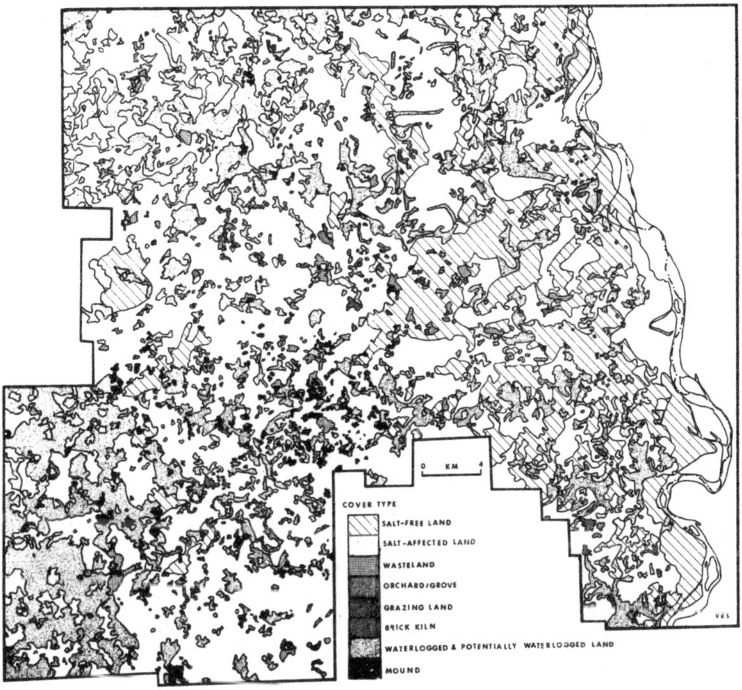


FIG. 2.2: Cover types in Sonipat region.

## Environmental Resource Units

### *Sub-Recent*

Remnants of the overflow channels are the only landforms of this period. Comprising of low-relief arcuate longitudinal bars with intervening broad shallow depressions. these overflow channels representing medium to coarse-grained sands were formed by the passage of floods through topographic lows along the rive bank. The photographic evidence suggests that flood routing during individual storm periods has occurred along the same course. Several embankments constructed west of these overflow channels have not only mitigated flood hazard, but also effectively controlled the encroachment of more productive lands by sandy accumulations. Thus, this environmental resource unit that covers some 70.64 sq km area or nearly 4.7 per cent of the total investigated area is a permanent feature of the geomorphic landscape.

TABLE 2.1: Area in Different Environmental Units in Sonipat Region

| Period | Environment units | Area in (sq km) | | | Total area (sq km) |
|---|---|---|---|---|---|
| | | Sonipat district | Rohtak district | Union territory Delhi | |
| Sub-Recent | Overflow-channels | 45.41 | – | 25.23 | 70.64 |
| Recent | Meander-cut plain | 1152.40 | 238.08 | 31.31 | 1421.79 |
| | Point bar | 13.73 | – | – | 13.73 |
| Pleisto-cene | Upland | 1.00 | – | – | 1.00 |
| Total | | 1212.54 | 238.08 | 56.54 | 1507.16 |

With limitations of grain texture, this unit possesses excellent surface drainage, and is generally free of salt hazards. Overflow channels are a suitable habitat for tall grass that trends to grow wildly throughout. The unit is ideally suited for dry-land farming of coarse foodgrains, and creeping corps.

*Recent*

Meander-cut plain and point bar accretions, the two major resource units of the Recent geological period, have been recognised. Covering 1421.79 sq km, the meander-cut plain is the single largest resource unit accounting for 94.3 per cent of the Sonipat area. It is morphologically segmented into meander scars, abandoned and buried channels, oxbows and broad meander necks. Villages like Bidhal and Pinana in Sonipat district, and in the northwest of the study area, are situated atop beautiful broad meander necks. By definition, meander neck is a topographic feature resulting from the process of neck cutoff in a freely meandering belt.

The photographic interpretation of the meander-cut plain suggests that besides swings of larger magnitude in the meander belt towards the east, the channel migration, especially of the Jamuna river, has also occurred through small intersecting meander loops. Whereas the former condition implies a major change in the river course, the latter activity suggests normal shifting of the meandering stream course. It is, therefore, reasonable to assume that periods of relative quiescence enabling the river to resume its normal activity have commonly occurred between periods of major channel displacements. The anomalously large area of the meander-cut plain may perhaps be explained by displacements in the subsurface faults a few thousand metres beneath the alluvial cover. It is perhaps in response to the neotectonic movements along these faults that the Jamuna has been able to rapidly change its course since the Quaternary (Sinha, 1986).

The micro-relief variations caused by numerous

form elements of the meander-cut plain have provided for a poor surface drainage. Consequently, salt-affected, waterlogged and potentially waterlogged conditions are intrinsic to the topographic characteristics of this unit. Notwithstanding the above geomorphic hazards, a thick cover of alluvium and fertile soils of the meander-cut plain provide ideal conditions for both intensive and extensive agricultural practices.

Point bars are in-channel deposits comprising of alternating series of arcuate to linear ridges of stratified sands to mud-sized material. The intervening depressions or swales are accumulations of finer sediments from overflows. In most cases, these intervening swales provide ideal conditions for waterlogging. Each point bar is a manifestation of channel migration during floods.

A small area, such as of 13.73 sq km in the point bar unit may probably be explained by the destruction of accretion topography during channel migration. This process has already been alluded to in reference to the meander-cut plain.

Sandy texture of material, large and rapid undulations, waterlogged conditions and usually small-size of the point bar unit do not normally permit its productive use. During the field check in 1987, it was, however, found that a part of the point bar topography near Murthal has been levelled down and brought under the plough.

### *Pleistocene*

The original surface of the Indo-Gangetic alluvial tract that has not been affected by fluvial activity is referred as the upland surface. Due to a widespread activity of laterally migrating streams, only 1.00 out of a total of 1507 sq km of the study area remains in this unit. The upland has an expressionless, monotonously flat surface of deep alluvium.

## Land Cover

High resolution remotely-sensed data on land cover (Table 2.2) and the thematic map on 1: 50,000 scale (Fig. 2.2) provide reliable land resource information useful for monitoring specific land resource conditions. The information so obtained is useful for land planning and management. Since most of the area is under cultivation, only the first order information, excluding urban and built-up area, has been mapped from aerial photographs. Also included in the eight cover types are areas under orchard/grove, and grazing land – the two other components of agricultural land-use. Besides, information on salt-free land, salt-affected land, waterlogged and potentially waterlogged land, wasteland including village common land, brick kiln and ancient habitation sites as mounds have also been interpreted.

Salt-free lands usually occupy well-drained higher surface in the overflow channel and parts of the meander-cut environmental resource units. These lands are generally confined to the eastern section, and occupy some 267.8 sq km or nearly 18.4 per cent of the investigated area.

Covering about 780.4 sq km or nearly 51.8 per cent of the study area, the salt-affected lands are more widespread than other types, and extend over parts of the meander-cut plain and the whole of the upland environmental resource units. These lands can be easily recognised on aerial photographs by the gray tone and slight upbevelling of land caused by the capillary movement of salt-rich subsoil water. According to the studies by Bhargava (1977, 1985) and Bhargava and Sharma (1984), these salt-affected lands are generally moderately to highly saline throughout the depth of soil profile, in which chlorides and sulphates of sodium and magnesium predominate.

TABLE 2.2: Area in Different Cover Types, Including Built-up Land in Sonipat Region

| Cover type | Area in (sq km) | | | Total | Percentage of Total area |
|---|---|---|---|---|---|
| | Sonipat district | Rohtak district | Union territory Delhi | | |
| Salt-free land | 258.80 | 2.76 | 15.21 | 276.77 | 18.36 |
| Salt-affected land | 600.05 | 149.61 | 30.72 | 780.38 | 51.79 |
| Wasteland, including village common land | 31.12 | 6.00 | 0.77 | 37.89 | 2.51 |
| Orchard/grove | 8.21 | 0.16 | 0.19 | 8.56 | 0.57 |
| Grazing land | 0.38 | 0.11 | — | 0.49 | 0.03 |
| Brick kiln | 4.68 | 0.38 | 1.19 | 6.25 | 0.41 |
| Waterlogged and potentially waterlogged land | 274.09 | 93.03 | 8.21 | 375.33 | 24.91 |
| Mound | 15.02 | 6.12 | 0.35 | 21.49 | 1.42 |
| Total | 1192.35 | 258.17 | 56.64 | 1507.16 | 100.00 |

Wasteland, including the village common land, is estimated at 37.9 sq km or 2.5 per cent of the total investigated area. A major part of this land is spread over the meander-cut plain resource unit. Following the definition given by the National Remote Sensing Agency (NRSA, 1985), wasteland is recognised as that land which is either not put to use or is being not used to its optimum potential due to some constraints. As shown in Table 2.3, all of this wasteland may be classified as culturable wasteland.

A culturable wasteland has potential for the development of agriculture, pastures and afforestation, but due to certain constraints it is not being used to its optimum potential. The principal cause of wasteland in the Sonipat region is the neglect due to excessive salts in soils. Abandoned brick kilns, most of which find association with abandoned or palaeo channels, rank a poor second in adding to the wasteland category. This area of about 5.3 sq km can be easily reclaimed with little economic input, and developed into some other uses of agricultural land.

Practically all of orchard/grove cover type is restricted to Sonipat district, where 8.2 of 8.5 sq km in this category is distributed. Further, within orchards and groves, the latter alone accounts for about 98 per cent of the area in the combined category.

Grazing lands are marginal and inferior in quality to foodgrain production. Only about 0.5 sq km area had been determined to be in this cover type.

Land in brick kiln includes only functional brick kilns. Most brick kilns, as noted previously, are site specific with reference to palaeo channels where loamy clay or clay loam in parts of abandoned channels is available for brick-making. In due course of time, this area of about 6 sq km will, however, be added to the category of culturable wasteland.

Waterlogged and potentially waterlogged land is closely associated with the meander-cut plain resource unit where topographic characteristics favour accumulation of water at or within 1.5 m depth of the surface, or where the ponded water drains very slowly. These lands cover nearly 25 per cent of the 1507 sq km area in the Sonipat region.

and are a common occurrence in canal command areas. Some recent newspaper reports suggest that steady rise of saline water-table since 1974 has lead to widespread waterlogging and saline water encroachment in the area under reference. Such areas are discernible on aerial photographs by a darker tone, and from their lower topographic position relative to the surrounding area.

TABLE 2.3: Classification of Wasteland in Sonipat Region

| Wasteland due to | Area (sq km) in | | | Total (sq km) | Percentage of total area |
|---|---|---|---|---|---|
| | Sonipat district | Rohtak district | Union territory Delhi | | |
| Neglect/Excessive Salts | 24.85 | 5.57 | 0.66 | 31.08 | 82.03 |
| Slope | 1.49 | — | — | 1.49 | 3.93 |
| Abandoned brick kilns | 4.78 | 0.43 | 0.11 | 5.32 | 14.04 |
| Total | 31.12 | 6.00 | 0.77 | 37.89 | 100.00 |

Mounds are probably relic of former settlements, and cover sizable area in the meander-cut plain of Sonipat and Rohtak districts. Closely spaced mounds, such as in the southwest of the region also provide for sufficient relief variations to impede surface drainage and, consequently, accentuate the effects of waterlogging.

## Agricultural Land-use

Three villages, one each from salt-free, salt-affected, and waterlogged and potentially waterlogged lands were randomly selected to the study of cropping pattern of the year of photography. These villages are: Jakhauli (S.O.I. map $53^{H}/1$) in the salt-free land; Salimsar Majra (S.O.I. map $53^{D}/13$) in the salt-affected land and, Khubrhu (S.O.I. map $53^{C}/16$) in the waterlogged and potentially waterlogged land. The purpose of this investigation is to integrate the information on land cover with agricultural land-use data for the purpose of suggesting alternative crops that can be most profitably raised to enhance the productivity of land within the limits of classified environment. In this approach, it is, however, implicit that the cropping pattern of any one of the above village applies uniformly to the various occurrences of the land cover to which it is referred.

*Jakhauli*

Maize and fodder are the dominant *kharif* crops of the village. Sugarcane and rice, the other crops of the season, are raised mainly for domestic consumption. Wheat is by far the most important commercial crop of the *rabi* season that accounts for more than 50 per cent of the cultivated area. At places, mustard is an intercrop with wheat. Besides, gram and fodder are also the favoured crops of this season. Vegetables are raised partly for commercial purpose, and during both the seasons. Since, maize, sugarcane, mustard, and gram are particularly salt sensitive crops, the cropping pattern of the village is in conformity with the soil environment.

The productivity level of this salt-free land can be enhanced by introducing oilseeds and groundnut, and citrus lemon fruit, all of which are ideally suited to cultivation on land that is free of salt hazards. Strengthening of irrigation facilities and the supporting agricultural inputs can also further the yield of sugarcane and gram under irrigated conditions.

*Salimsar Majra*

With as many as eleven crops grown, the cropping pattern in the village may be described as diverse. Sugarcane competing for the cultivated area but lagging far behind bajra and jowar in area commanded, the *kharif* cropping pattern is largely dominated by coarse foodgrain crops. Wheat, gram, and fodder represent the decreasing order of importance of the *rabi* season crops. Cotton, the perennial crop is cultivated on farms of 4 hectare or more in size.

The cultivation of salt-sensitive crops in Salimsar Majra is rather difficult to explain in the present context. Either this land suffers from low to very low soil salinity hazards or there may have been a few isolated parcels of salt-free land that could not be represented on the mapped scale due to their small size. The soil salinity appears to have intensified over the years, as is evident from the disappearance of jowar and gram – the more salt-sensitive crops. The 1986-87 cropping pattern of rice-wheat-fodder-sugarcane-pulses in Salimsar Majra probably shows a better adjustment with the salt-affected conditions of cultivated land.

The productivity of these lands can be enhanced by introducing crops that have medium to high tolerance to salts. Some of these crops are *dhaincha*, barley, *barseem*, paddy, wheat, and cotton. Alternatively, the salt-affected lands can be reclaimed by the application of gypsum with adequate irrigation.

*Khubrhu*

Khubrhu represents an environment of very high watertable conditions. In 1969-70, the village raised fodder, *bajra* and sugarcane in *kharif*, and wheat and gram in the *rabi*. In this case also, the above crops do not relate well higher watertable generally of marginal and saline quality. That the cropping pattern in 1986-87 revolved around rice, fodder, sugarcane, pulses and wheat suggests further rise of watertable over a period of 17 years. Being an unstable and dynamic environment, the waterlogged and potentially waterlogged lands require special agronomic and engineering practices to upgrade their productivity levels. The terrain characteristics of this land, as noted previously, cannot be modified easily. Hence, development of vertical and lateral drainage from the deep subsoil should be considered to mitigate the hazards.

## Dynamics of Salt Hazards

Analyses of a selected few physico-chemical properties of soil sample given in Table 2.4 roughly conforms to the pattern of cover types in the Sonipat region. The determination of other chemical properties of soils, such as saturation extracts and exchangeable cations including SAR and ESP, through desirable, could not be carried out for want of adequate laboratory facilities.

Soil samples 2 and 3 possess pH from 7.8 to 8.1 and EC from 0.49 to 1.12 mmhos/cm at 25°C. These values indicate a salt-free environment as interpreted on the 1969 aerial photographs. The soil with salt-hazards show pH in the range of 8.4 to 9.5, and EC between 0.68 and 4.99. These lands generally tends to be saline in nature. The pH of the soils of the potentially waterlogged environment ranges from 7.8 to 9.2, and EC from 0.65 to 4.92 mmhos/cm at 25°C. These soils may perhaps be referred to as alkali in nature.

TABLE 2.4: Description and Analyses of Selected Physio-Chemical Properties of Soil Samples from Sonipat Region

| Sample number | Site | Photointerpretation* | Sample from | Other description | Physico-chemical properties | | |
|---|---|---|---|---|---|---|---|
| | | | | | Soil field texture | pH (1:2) solution | EC m mmhos cm at 25°C |
| 1. | Intersection of road and Samalkha Minor | III | Wheat field | *Sarson* | Loam | 8.4 | 0.94 |
| 2. | Between Pabnera and Ghasauli villages | I | Sugarcane field | Fodder, oilseed marked micro-relief variations | Loam | 7.8 | 1.12 |
| 3. | East of Pabnera Bund | I | Fallow land | Wheat, sugarcane, oil-seeds and vegetables | Sandy loam | 8.1 | 0.49 |
| 4. | Mimarpur village | II | Potatoe field | Wheat and *Sar-son* intercropped, vegetables | Loamy sand | 8.4 | 0.68 |
| 5. | Point bar west of Murthal and across Rajpur Distributary | III | Wasteland | Rapid and large relief undulations | Fine sand | 8.4 | 0.65 |
| 6. | Point bar east of Bahalgarh | II | Barren | Vast stretch of land showing surface efflo-rescence covered with thistle and beri shrubs | Silty clay | 9.5 | 4.99 |
| 7. | West of Western Jamuna Canal | II | Fallow land | Depessional land | Silty clay | 8.8 | 0.91 |
| 8. | West of Kharkhauda village | II | Wheat field | Slightly elevated land, stunted crop growth | Loam | 8.4 | 0.77 |
| 9. | South of Kharkhauda village | II | Wheat field | Stunted crop growth | Loam | 8.7 | 2.25 |
| 10. | Sehari village | III | Fallow land | — | Silty clay | 9.2 | 4.92 |
| 11. | Bidhal village | III | Wheat field | Meander-neck slope | Loam | 7.8 | 2.45 |

* I - Salt-free land; II - Salt-affected land; and III - Waterlogged and potentially waterlogged land as interpreted on aerial photographs.

The low electrical conductivity (EC) observed in the most of the samples is not commensurate with the relatively higher values reported by Bhargava and Sharma (1984) and Bhargava (1985). The discrepancy may be due to the fact that in the drought year of 1987, a very heavy night-time downpour had occurred a day before the samples were collected from the region. This substantial rainfall could have easily leached the salts from the root depths to the lower soil horizons. It is also remotely possible that the plots from which soils were collected have been reclaimed from the salt-hazards. The above tabulated EC values are, therefore, not indicative of the 1969 soil environment, or of the chemical composition of the soils under normal conditions.

Boreholes extending the depths between 49 and 207 metres at Rattangarh, Bhatgaon and Kharkhauda in Sonipat district, and Kansala and Sampla in Rohtak district were drilled by the Haryana State Minor Irrigation Tubewells Corporation, Karnal for appraisal of aquifers and groundwater characteristics. These data presented in Table 2.5 have been interpreted for a tentative explanation of the geological environment of alluvium, and of geohydrological characteristics of the aquifer for explaining the occurrence of salt-affected lands. Three cycles of deposition at Kansala, two each at Bhatgaon and Kharkhauda, and one each at Rattangarh and Sampla may be inferred from the litholog data. It is interesting to note that marine horizons are absent in these, and the other lithologs available in Haryana. The cyclic sedimentation, therefore, tentatively suggests either a subsiding through or a change in the river regime that has deposited the alluvium. Further investigation into this aspect of foredeep deposition is, however, outside the purview of this research. Relevant to the explanation of salt-affected lands is the observation of *kankar* found dispersed to great depths in the subsoil, and in horizons comprising of fine to medium sands and clays. These secondary concretions are mostly carbonates of calcium formed in an environment of shallow and fluctuating water table conditions (Chadha, 1983; Hardie, 1984). Chadha (1983) also noted that the salinity of groundwater is closely related to the aquifers associated with *kankar*. Whereas the groundwater is generally fresh between the Jamuna river and the National Highway No. 1, its quality deteriorates westward (Gupta, 1983a). Further, the watertable in saline tracts of Sonipat and Rohtak districts in between 1 and 3 metres from the surface (Gupta 1983b).

TABLE 2.5: Borehole Data from Selected Sites
Village - Rattangarh (Sonipat district)

| Lithology (Field description) | Depth range (m) | Thickness (m) |
|---|---|---|
| 1 | 2 | 3 |
| Clay, brownish green | 0-6.10 | 6.10 |
| Silt, brownish green | 6.10-12.19 | 6.09 |
| Clay, brownish green | 12.19-15.24 | 3.05 |
| Silt, brownish green, mixed with very find sand | 15.24-18.29 | 3.05 |
| Clay, brownish grey, with kankar | 18.29-24.38 | 6.09 |
| Fine sand, brownish black | 24.38-36.58 | 12.20 |
| Fine sand, brownish, micaceous | 36.58-48.77 | 12.15 |
| *Village - Bhatgaon (Sonipat district)* | | |
| Top soil, silt | 0-1.83 | 1.83 |
| Very fine sand, brown | 1.83-7.01 | 5.18 |
| Silt | 7.01-13.10 | 6.09 |
| Sandy clay | 13.10-25.30 | 12.20 |
| Fine to medium sand, light brown | 25.30-31.39 | 6.09 |

*Contd.*

| | | |
|---|---|---|
| Clay, mixed with medium to coarse sand | 31.39-37.49 | 6.10 |
| Sandy clay | 37.49-43.59 | 6.10 |
| Fine sand, mixed with clay | 43.59-48.77 | 5.18 |
| Fine sand, grey, micaceous | 48.77-61.87 | 13.10 |
| *Village - Kharkhauda (Sonipat district)* | | |
| Clay, pale brown, with silt | 0-1.52 | 1.52 |
| Fine sand, brown, mixed with silt | 1.52-18.29 | 16.77 |
| Clay, brown, with *kankar* | 18.29-21.33 | 3.04 |
| Medium to coarse sand, brown, *kankar* | 21.33-24.99 | 3.66 |
| Medium sand, dark grey, mixed with quartz and mica | 24.99-27.43 | 2.44 |
| Fine to medium sand, light grey, mixed with quartz and mica | 27.43-30.48 | 3.05 |
| Clay, pale brown, with *kankar* | 30.48-48.77 | 18.29 |
| Medium to coarse sand, greyish brown | 48.77-55.78 | 7.01 |
| Clay, with crushed *kankar* | 55.78-71.93 | 16.15 |
| Medium sand, dark grey | 71.93-73.76 | 1.83 |
| Clay, pale brown, mixed with *kankar* | 73.76-96.62 | 22.86 |
| Very fine sand, reddish brown, mixed with silt | 96.62-102.41 | 5.79 |
| Clay, brown, mixed with *kankar* | 102.41-134.11 | 31.70 |
| Clay, pale brown | 134.11-152.40 | 18.29 |
| *Village - Kansala (Rohtak district)* | | |
| Fine sand, greyish brown | 0-3.05 | 3.05 |
| Fine to medium sand, greyish colour | 3.05-6.10 | 3.05 |
| Fine sand, brownish black | 6.10-9.14 | 3.04 |
| Fine sand, greyish, with mica flakes | 9.14-15.24 | 6.10 |
| Fine sand, greyish | 15.24-18.29 | 3.05 |
| Fine sand with silt, greyish, with *kankar* | 18.29-24.38 | 6.09 |
| Fine to medium sand, greyish, with mica flakes | 24.38-27.43 | 3.05 |
| Medium to coarse sandy *silt*, greyish | 27.43-30.48 | 3.05 |
| Fine to medium sand, brown, with mica flakes | 30.48-35.05 | 4.57 |
| Fine sand, greyish with few *kankar* | 35.05-36.58 | 1.53 |
| Fine sand, greyish | 36.58-48.77 | 12.19 |
| Fine sand, greyish brown, crushed *kankar* | 48.77-54.86 | 6.09 |
| Fine sand with clay, greyish brown, crushed *kankar* | 54.86-60.96 | 6.10 |
| Fine sand, reddish brown, with crushed *kankar* | 60.96-64.00 | 3.04 |
| Clay, grey brown, with crushed *kankar* | 64.00-70.10 | 6.10 |
| Fine sand, reddish | 70.10-73.15 | 3.05 |
| Fine to medium sand, reddish brown | 73.15-76.20 | 3.05 |
| Fine sand, reddish brown, with crushed *kankar* | 76.20-79.25 | 3.05 |
| Fine sand, grey, with mica flakes | 79.25-85.34 | 6.09 |

*(Contd.)*

| 1 | 2 | 3 |
|---|---|---|
| Fine sand, grey, with crushed *kankar* | 85.34-128.02 | 42.68 |
| Clay with sand, with crushed *kankar* | 128.02-140.21 | 12.19 |
| Fine sand, with *kankar* | 140.21-146.30 | 6.09 |
| Clay, with about 70% *kankar* | 146.30-153.31 | 7.01 |
| *Village - Sampla (Rohtak district)* | | |
| Top soil, yellowish brown | 0-3.05 | 3.05 |
| Clay, greyish brown with *kankar* | 3.05-65.23 | 62.18 |
| Calcareous nodules, brownish grey, mixed with fine sand and silt | 65.23-67.05 | 1.82 |
| Silt, brownish black | 67.05-70.10 | 3.05 |
| Clay, greyish brown, mixed with *kankar* | 70.10-115.82 | 45.72 |
| Silt, brownish grey, with clay and *kankar* | 115.82-121.92 | 6.10 |
| Very fine sand, yellowish grey, mixed with *kankar* | 121.92-124.97 | 3.05 |
| Clay, yellowish grey | 124.97-131.06 | 6.09 |
| Silt, with *kankar* | 131.06-140.21 | 9.15 |
| Clay, with *kankar* | 140.21-158.49 | 18.28 |
| Silt, with *kankar* | 158.49-167.64 | 9.19 |
| Clay, with *kankar* | 167.64-176.78 | 9.14 |
| Silt, with *kankar* | 176.78-192.02 | 15.24 |
| *Kankar* | 192.02-195.07 | 3.05 |
| Sticky clay | 195.07-198.12 | 3.05 |
| Kankar, with clay | 198.12-201.17 | 3.05 |
| Silt, with clay and *kankar* | 201.17-207.26 | 6.09 |

Source: Haryana State Minor Irrigation Tubewells Corporation, Karnal.

The dynamics of saline soils has been discussed in detail by Bhargava and Sharma (1984). Accordingly, salt effects are perpetuated by high salt-content and shallow depth of groundwater, stagnation of water at the surface, extensive application of canal and minor irrigation with resulting seepage, poor drainage from irrigated lands, and high potential evapotranspiratory demands throughout the year. Field studies by these and other scientists suggest that the watertable in this area fluctuates between 0.8 and 1.5 metres from the surface. It has been further observed that the critical depth of groundwater from where the salts are transported towards the surface by capillary action is less than 1 m. Such watertable conditions are common during and immediately after the monsoon season. Field data also suggest that salts can also be transported from a soil that is saturated at a depth of 55 to 150 cm from the surface. Such saturated soil conditions are common under irrigated agriculture and its associated infrastructure. Salt hazards adversely affect crop yields, and can render much of the valuable land resource into wasteland. Provision of an efficient system of subsurface drainage is perhaps a good solution for mitigating the salt hazards.

The alkali soils represent excessively wet or poorly drained lands situated below higher ground. These low-lying tracts receive surface runoff from the surrounding area, and accumulate in the sodic environment deflocculated clay particles and silicate minerals. The shallow fluctuating watertable also encourages the precipitation of the carbonates of calcium and magnesium. The alkali soils are rich in calcium carbonate content.

# 3

# LAND RESOURCES AND LAND RESOURCE PLANNING IN REWARI REGION

## Introduction

COMPRISING OF the parts of Rohtak, Gurgaon, Mahendragarh and Bhiwani districts, the Rewari region covers an area of 2887.75 sq km between latitudes 28°00' and 28° 30' north, and longitude, 76°15' and 77°00' east (Fig. 3.1). Ateli, Nahar, Khol, Rewari, Bawal, Pataudа, Pataudi, Farruknagar, and the upcoming industrial town of Dharuhera are some of the important settlements of the region.

According to Troll's classification, the Rewari region has a dry tropical climate in which, to quote Rani (1978), protracted periods of dry-sequence are typical of the climatic regime. The average, July through September, summer-season rainfall at Rewari and Farruknagar, reported to be 39.1 and 36.7 cm respectively, is nearly 76 per cent of the annual rainfall total recorded at these two raingauge stations. This summer rainfall at Rewari is experienced in 20.9 and that at Farruknagar in 16.6 rainy-days (Sharma, 1980). Similarly, 3.6 and 2.9 cm of December through February rainfall dispersed respectively over 6.9 and 6.1 rainy-days is obtained at Rewari and Farruknagar. The meagre rainfall and high temperatures throughout the year cause perpetual drought and crop failures in the study area. The existing deep and unusable groundwater resources over a large part of the Rewari region, and surface water potential of only a limited number of flashy inland drainage systems cannot be effectively harnessed for furtherance of agricultural pursuits without deteriorating the quality of environment.

The existing water supply over most of the study area is inadequate to meet the water requirement of crops. Cultivation under such conditions involves high economic risk. Dryland environments worldover are fragile and ecological risk areas on account of the paucity of water. In the context of resource planning, such an ecological system is, to quote Shankaranarayan (1985), so delicately adjusted to the scarcity of water that it cannot accommodate any external biotic disturbance. Further, under pressure of human use, the dryland environments also lose productivity and recuperative ability. Thus, cultivation on sand dunes and marginal lands, non-judicious planning and application of irrigation water, uncontrolled and overgrazing, and indiscriminate felling of whatever trees that grow cause extensive degradation of the environment leading to a host of environmental problems, such as droughts, waterlogging, growth of soil salts, soil erosion and deterioration of soil fertility.

Sahibi and Krishnawati (Kasauti) are the two inland streams of the region, and sources of substantial recharge of fresh-quality water into groundwater which otherwise is brackish. A barrage across Sahibi has been constructed at Masani near Rewari for providing flood protection and utilisation of stored water for *kharif* crops. In the study area,

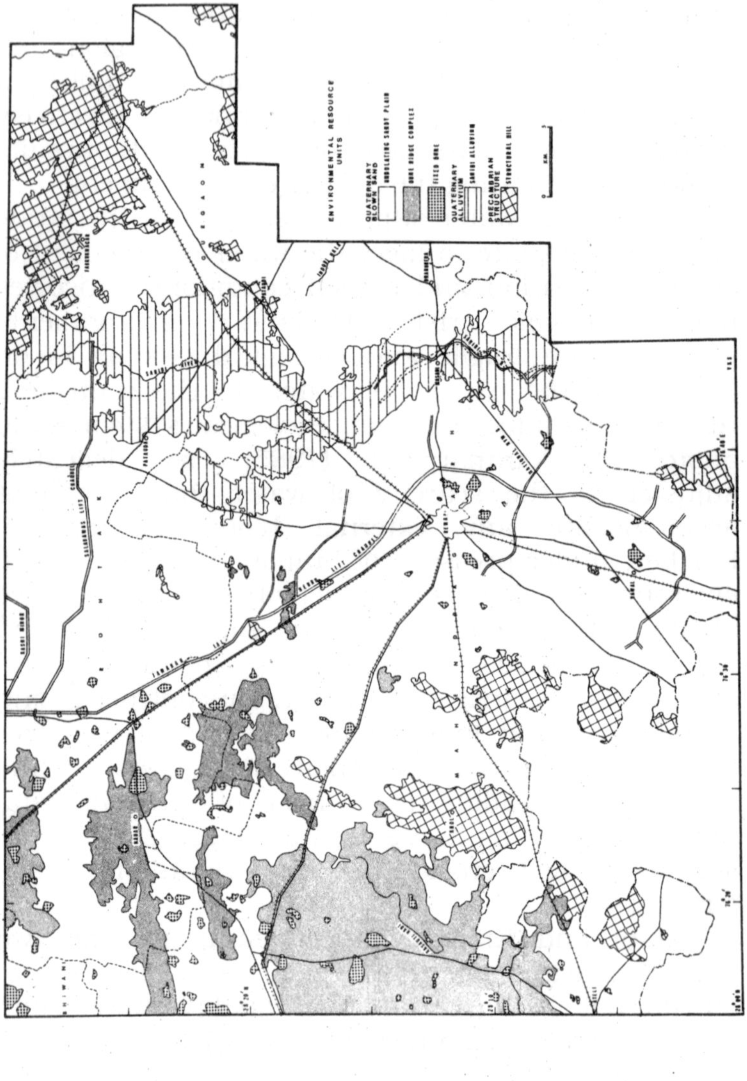


FIG. 3.1: Classification of dryland environment and associated landforms in Rewari region.

groundwater recharge due to the other stream has now been checked following construction of small dams on Krishnawati in Rajasthan. Consequently, the watertable along the river courses in the study area has fallen considerably. Here, the replacement of fresh quality groundwater by brackish and saline subsurface water is of special significance to the degradation of agro-environment and to rural development. A few torrents also originate on slopes of the hills in the study region, but they make short channels that peter-out on the adjacent sandy covers.

In view of the scarcity of water resources for sustained agricultural growth, several lifts irrigation schemes in southwestern Haryana have been pressed into service. The study area is served by the Jawaharlal Nehru (JLN) lift irrigation scheme. The JLN canal takes-off from the Delhi Parallel Branch, but the system will become fully operational on the completion of Sutlej-Yamuna Link in Punjab. The JLN canal and most of its distributaries have been lined for minimising conveyance losses. The project has been designed for 24 per cent intensity of irrigation in the *kharif*, and 38 per cent in the *rabi* cropping season.

## Geology

The Rewari region is the domain of dryland landscape of largely the Quaternary blown sand and alluvium of endohric streams, and the Precambrian Aravalli hill outcrops. The aeolian sand represents a variety of form from dune complex, dunes stabilised by vegetation, sand shadows, and vast stretches of undulating sandy plain. Goyal and Singh (1982) also report fossil or relic dunes in part of the Rohtak district studied in this research. However, the photographic interpretation does not support the morphologic evidence of fossil dunes (see, Guide *et al.*, 1973) from anywhere in the study area. Alluvium is spread over a wide tract along the Sahibi. This alluvium has been deposited largely from storm-period runoff. Localised accretion of alluvium has also been observed along the course of Krishnawati stream, and at the foot of hills. Aravalli hills are represented by the quartzites of Alwar series in Rohtak district, and by the folded shale and slate with quartzite incalcations, quartzites, biotite schists, gneisses, calcareous ampheboles and marbles of Ajabgarh series, also of the Delhi system. Another Archaean Dharwar System of rocks in Narnaul and Mahendragarh tahsils comprises of crystalline limestones, shales, quartzites, slates, gneisses and sandstones. Some of these rock formations are associated with semi-precious stones, and important minerals like limestone, flexible red sandstone, slate, and ornamental marble. The exploitation of these minerals has provided the basis of mining industry, though on a small-scale. Further geological investigations in this region are required to fully explore and develop the mineral potential of the Aravalli hills. The development of mining industry in this region will provide employment to the local inhabitants and, thereby, ease the pressure on land resources for agriculture and grazing activity.

## Environmental Resource Units

The Quaternary blown sand and alluvium and Precambrian structural relief features, divisible into five environmental resource units, have been photo-interpreted in the Rewari region. Of these, the Quaternary blown sand alone covers some 83 per cent of the region to be followed by the Quaternary Sahibi alluvium with some 9 per cent of the total surveyed area. The remaining area is in the Precambrian structural relief. The districtwise area in each of the five environmental resource units has been summarised in Table 3.1.

The Quaternary blown sand is represented by three environmental resource units as undulating sandy plain, dune ridge complex, and fixed (stable) dune. The undulating sandy plain with nearly 82.33 per cent of the area in this environment is by far the most widespread resource unit. Due to its wide expanse and apparently uneven thickness of sand cover punctuated by sandy ridges, wide shallow hollows (*tals*), rocky protuberances and fixed dunes, the undulating sandy plain is one of diverse relief

TABLE 3.1: Area in Different Environmental Units in Rewari Region

| Period Process | Environmental Unit | Area in (sq km) | | | | Total area (km² |
|---|---|---|---|---|---|---|
| | | Bhiwani district | Rohtak district | Gurgaon district | Mahendragarh district | |
| Quaternary Blown sand | Undulating sandy plain | 25.95 | 484.11 | 374.01 | 1090.75 | 1974.82 |
| | Dune ridge complex | 0.97 | 96.19 | – | 277.77 | 374.93 |
| | Fixed dune | 2.59 | 14.92 | – | 31.39 | 48.90 |
| Quaternary Alluvium | Sahibi alluvium | – | 47.05 | 118.63 | 102.98 | 268.66 |
| Precambrian Structure | Structural hill | – | 19.82 | 97.74 | 102.88 | 220.44 |
| Total | | 29.51 | 662.09 | 590.38 | 1605.77 | 2887.75 |

composition. At places these features provide a relief of up to 7 m within the environmental resource unit. In general, the elevation of the undulating sandy plain varies from about 300 m in the southwest to 218 m in the north and 237 m in the east. This elevation variation is in conformity with the northward directed slope due to the Aravalli hills which eventually culminates in the topographic sag in Rohtak district.

The dune ridge complex environmental unit comprises generally of east-west oriented transverse sandy ridges which also display forked, parabolic, and complexly intersecting pattern of ridges. The asymmetrical ridges are irregularly spaced and are of variable lengths. The elevation in the various occurrences of the environmental unit varies from about 280 m in the south to about 225 m in the north. The ridges are generally 5 to 7 m high, but a relief of some 15 m is not uncommon for longer and broad-based transverse dunes. Slopes of ridges are under rainfed cultivation, which is now supported by sprinkler irrigation. Transverse dunes develop in areas where the sandflow rate is high, and abundant sand for dune building activity is provided by unconsolidated materials and rockbed. Absence or near absence of vegetal cover is another precondition for the formation of transverse dunes.

Fixed or stable dunes occur as isolated sandy hummocks. These dunes owe their stability to the growth of trees, grasses and weeds and shrubs, the common among which respectively are: *jaal* (*Salvadora oleoides*), *jand* (*Prosopis spicigera*) and *kikar* (*Acacia arabica*); *ak* (*Calotropis procera*), *pohli* (*Carthamus oxycantha*) and *tandla* (*Digera arvensis*); and *dhatura* (*Datura fastuosa*), *karir* (*Capparis aphylla*) and *jharberi* (*Zizyphus numularia*).

The alluvial tract formed by the endohric Sahibi, Krishnawati, and numerous other small hill torrents draining the slopes of the Aravalli hills is an important land resource of the region. The Quaternary alluvium deposited by the Sahibi floods is particularly significant as it covers 268.66 sq km area in parts of Mahendragarh, Gurgaon and Rohtak districts.

The furies of Sahibi river have brought periodic miseries to these areas. As described in Table 3.2, the Sahibi flood events of moderate to high magnitude have occurred once in 9 years from 1845 to 1909. The instantaneous peak flows from 1965 to 1984 observed at Masani Barrage (Table 3.3), and analysed for flood frequency analysis using Gumbel's method of extreme value analysis, suggest a recurrence interval of 3.1 years for the bankful stage of 350 cumec (Parmar, 1986). The peak flow of 3028 $m^3/s$ in 1977 is clearly an outlier that does not, however, fit into the general pattern of floods through the recorded period of flow. It was during the storm-flow of 1977 that the river breached the National Highway 8 near Khaliawas village; and developed a new channel in a short span of time.

The Sahibi floods described between 1845 and 1909 suggest that the area around Rewari town had been the target of flood fury. The photographic interpretation however fails to support the above observation. It may be due to sudden escape of moderate to high magnitude overbank floods from shallow broad channel. In this research, the Sahibi alluvium has been recognised on the basis of break of slope along river margins, marking the former banks of the channel, and discontinuous channels of the Sahibi that are so prominent some distance from the present-day river course.

Elevation in the alluvial tract of the Sahibi river varies from about 249 m in the southeast to about 220 m in the north. A relative relief of some 2 m due to the presence of rocky protuberances within the alluvium has also been observed. The area of Sahibi alluvium constitutes one of the many groundwater basins in Haryana, and is known for fresh-quality groundwater reservoir.

TABLE 3.2: Description of Historic Floods in the Sahibi River (1845 to 1909)

| Sl.No. | Year | Description |
|---|---|---|
| 1. | 1845 | The town of Rewari was engulfed by floods after a heavy rain occurred in the Jaipur district of Rajasthan. The Sahibi in Alwar district, Rajasthan is said to have abandoned its long established channel. |
| 2. | 1883 | Rewari was devastated in August 1883 by a flood that was nearly 1 m deep in the town. Though the flood drained northward within a short period of 3 to 4 hours, it had in this short time caused extensive damage within and outside the town by destroying, among others, the railway bunglow and a large portion of the railway tract. |
| 3. | 1885 | A peak discharge in the Sahibi caused extensive deposition of a thick cover of sand, particularly along the railway tract where the flood waters were temporarily ponded. |
| 4. | 1887 | This flood in Rewari area was as severe as the one in 1885. |
| 5. | 1888 | A flood of a serious nature created several new flood channels through which the waters continued to drain for about 5 days. |
| 6. | 1908 and 1909 | These high magnitude floods recurred after a lapse of nearly 20 years. The flood waters occupied vast areas, and took nearly three weeks to drain. |

Source: Government of Punjab, 1916. *Punjab District Gazetteer*, Vol. IIA. Hissar District (Part A), Lahore, Punjab.

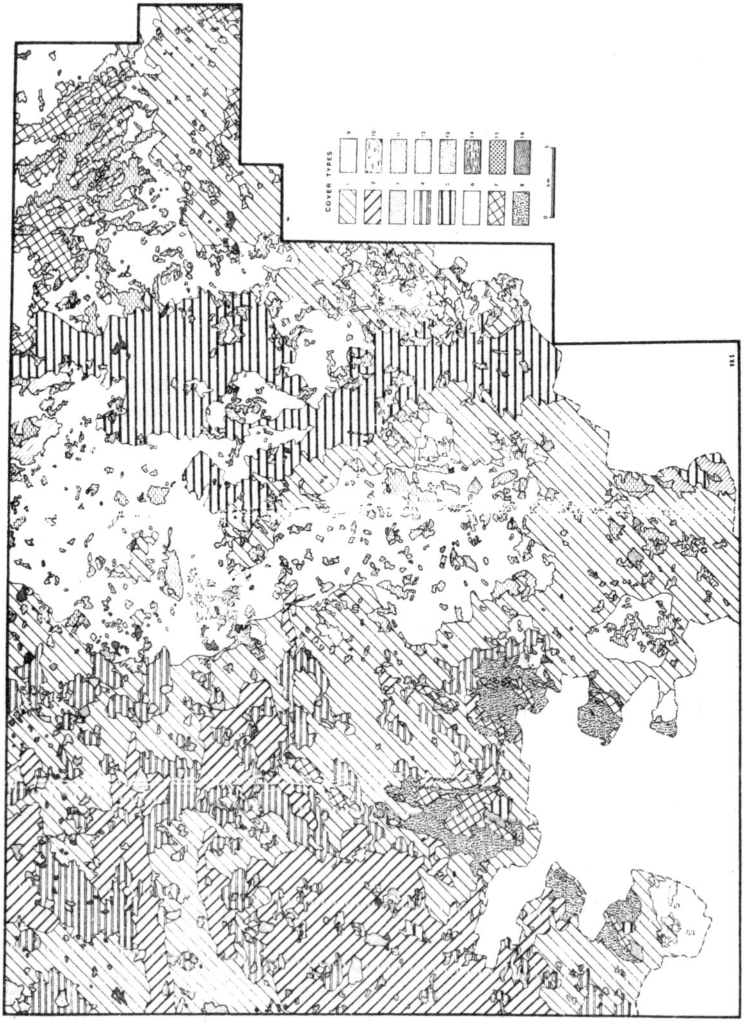


FIG. 3.2: Cover types in Rewari region. 1. Sandy plain, 2. Dune ridge, 3. Fixed dune, 4. *Tal*, 5.Alluvium, 6. Salt-affected land, 7. Hill, 8. Pediment, 9. Broken land, 10. Marsh, 11. Waterlogged and potentially waterlogged land, including depressions, 12. Rocky protuberance, 13. Wasteland, 14. Brick kiln, 15. Orchard/grove, 16. Grazing land.

TABLE 3.3: Instantaneous Peak Discharges at Masani

| Year | Month | Date | Discharge ($m^3/s$) |
|---|---|---|---|
| 1965 | September | 4 | 451.52 |
| 1966 | August | 11 | 35.23 |
| 1967 | July | 4 | 254.70 |
| 1968 | July | 14 | 231.03 |
| 1969 | September | 11 | 258.05 |
| 1970 | September | 24 | 142.07 |
| 1971 | August | 16 | 140.79 |
| 1972 | August | 12 | 189.09 |
| 1973 | September | 13 | 139.57 |
| 1974 | August | 4 | 434.12 |
| 1975 | August | 21 | 433.09 |
| 1976 | August | 19 | 675.13 |
| 1977 | August | 5 | 3028.00 |
| 1978 | September | 2 | 827.00 |
| 1979 | July | 16 | 84.11 |
| 1980 | August | 3 | 268.86 |
| 1981 | August | 6 | 216.32 |
| 1982 | August | 12 | 40.39 |
| 1983 | July | 28 | 119.54 |
| 1984 | September | 5 | 60.35 |

Source: S.D.O., Irrigation, Masani Barrage Project, Rewari.

The structural hill environmental resource unit represents a varied morphologic character of the outliers in respect of extent. Some outliers steeply rise from their base, while a few others make wide pediment slopes. In general, the hills near Khol in the southwest of the study area reach 472 m above the mean sea level, and rise to about 120 m from their surroundings. The elevation and relative relief of the hills gradually decreases towards the north and northeast. In the area around Farruknagar in the northeast, these outliers become subdued. They lie at about 220 m above the mean sea level in which the magnitude of relief variation is of the order of 5 to 20 m. It is in this area that the rolling hills are interspersed with water bodies. Elsewhere in the Rewari region, nearly all the west-facing slopes of hills are deeply dissected by gullies. Slate and limestone quarrying in the Precambrian structural hill unit is an important economic activity.

## Cover Types

Cover types associated with each of the five environmental resource units are classified in Table 3.4, and their spatial pattern depicted in Fig. 3.2. Land cover areas too small for representation at the mapped scale have, however, been omitted. Selected properties of only a few important cover types are discussed in this section.

The undulating sandy plain resource unit comprises of 9 cover types, of which the sandy plain with nearly 45 per cent of the area is by far the most important cover type of this unit. According to a classification by CAD (October 1986), the soils of sandy plain are sandy to sandy loam in texture, and possess single to a weak subangular blocky structure. The non-calcareous type of soils are referred to as mixed, hyperthermic, typic torripsamments and udic ustochrepts. The soils that tend to become calcareous with depth are referred to as typic ustochrepts.

Most of the sandy plain, to refer to the CAD (October 1986) report, is in Class-III and Class-V land capability classes. The Class-III lands occasionally possess moderately steep slopes, sandy soils with low moisture holding capacity, and low inherent soil fertility. The productivity level of these soils can be enhanced by adopting suitable soil conserving crop rotation that can promote high yields of cultivated crops, and by maintaining a satisfactory status of soils which are basically low in organic and nitrogen nutrients. The Class-V land is reportedly unsuitable for cultivation, but is suitable for grazing and forestry. The agricultural density is substantially high even in these poor productivity lands which leave no alternative choice for the farmers but to till the land for crop cultivation.

TABLE 3.4: Area in Different Cover Types, Including Built-up Land in Rewari Region

| Environmental Resource Unit | Cover Type | Area (sq km) |
|---|---|---|
| 1 | 2 | 3 |
| Undulating sandy plain | Sandy plain | 894.64 |
| | *Tal* | 193.17 |
| | Depression, waterlogged and potentially waterlogged land | 102.65 |
| | Salt-affected land | 631.81 |
| | Rocky protuberance | 68.78 |
| | Wasteland | 64.41 |
| | Brick kiln | 3.02 |
| | Orchard/grove | 2.51 |
| | Grazing land | 13.83 |
| Dune ridge complex | Dunal ridge | 350.58 |
| | *Tal* | 15.14 |
| | Rocky protuberance | 4.28 |
| | Alluvium | 1.72 |
| | Depression, waterlogged and potentially waterlogged land | 0.19 |
| | Wasteland | 3.02 |
| Fixed dune | Stable dune | 48.90 |
| Sahibi alluvium | Alluvium | 248.92 |
| | Marsh | 5.50 |
| | Waterlogged and potentially waterlogged land | 3.87 |
| | Salt-affected land | 0.66 |
| | Wasteland | 5.47 |
| | Rocky protuberance | 2.35 |
| | Orchard/grove | 1.79 |
| | Grazing land | 0.10 |
| Structural hill | Hill | 89.13 |
| | Pediment | 68.90 |
| | Broken land | 2.43 |
| | Depression, waterlogged and potentially waterlogged land | 44.32 |
| | Alluvium | 3.44 |
| | Salt-affected land | 11.53 |
| | Orchard/grove | 0.69 |

*Tals* or hollows within the undulating sandy plain cover some 9.78 per cent of the area in the undulating sandy plain resources unit. These extensively cultivated and irrigated lands possess loamy sand to coarse loam soil textures which probably refer to the Bhambewa series of Goyal and Singh (1982) and to the Lukhi series of CAD (October 1986). Designated as typic ustochrepts, these soils are free of salt-hazards. The limitations of these soils for the support of agriculture are, however, identical to the soils developed on blown sands. The *tals* suffer from additional disadvantage in that the centripetal slope and perhaps shallow depth to rock bed have lead to potentially waterlogged conditions in many occurrences of the cover type.

Waterlogged and potentially waterlogged lands, including depressions, cover another 5.24 per cent of the area. They find expression along canal network and within natural depressions due to rocky protuberances. It may be noted that surplus water from the Western Jamuna Canal system were being diverted by 1979 through the JLN lift irrigation channel for irrigation in the study area. The salt-affected lands covering 32 per cent of the area in this environmental resource unit also appear causally related with the recent developments in surface and subsurface water applications for furtherance of agricultural pursuits. Consequently, the soils suffer from very slight to moderately low alkali hazards. Here, the coarse loamy textured soils belong to the hyperthermic family of typic ustochrepts. Most of the salt-affected land is designated a Class-III capability status (see, CAD, October 1986). The land suffers from excessive wetness and sodicity, sandy soils of low field capacity, and low inherent fertility. On much of the nearly level land of this class, provision of subsurface drainage, and a cropping plan of deep-rooted legumes grown in rotation are required. In some irrigated areas, high watertable, slow permeability and accumulation of salts further limit the productivity potential of Class-III lands.

Grazing lands, the marginal quality productivity resource unit, spread over 0.70 per cent of the environmental unit under reference. In this chapter, as elsewhere in this research, grazing lands are defined by areas of slight relief on usually hummocky surfaces with a depression within, and covered with some form of vegetation usually of short, sparsely dense trees or bushes and shrubs. Sometime these lands also show evidence of excessive salts. Presence of a depression is, however, considered diagnostic of grazing land, provided relief and vegetation characteristics are similar to the one referred to above.

The dune ridge complex is pre-eminently dominated by dune ridges, and five other cover types. These ridges in part of the Rohtak district included in this chapter are referred to by Goyal and Singh (1982) as fossil dunes. In this case, the available soil classifications of the area do not seem to indicate a difference in the physico-chemical properties of the soils of the dune ridges when compared with the other environmental resource units of the blown sand. A part of this covers-type is classified by CAD (October 1986) in Class-IV land. Soils of this capability class can be best maintained in perennial vegetation, but may be cultivated for certain specialised crops that require special land management techniques. Loose sand grains susceptible to wind deflation and magnitude of slope are parameters adversely affecting the agricultural use of dune ridges that are inherently unproductive. *Tals* within the dune ridge complex offer better conditions for utilising the resource potential of the area, but the extent of such areas is limited. Alluvium deposited by the inland Krishnawati before it looses itself completely in the dune ridge complex is a relatively important resource base within the above mentioned environmental resource unit. Relatively flat surface, sandy loam to loamy sand fluviatile soils mixed with blown sand, and fresh-quality seasonally high watertable are the properties of the alluvium denied to other cover types in the dune ridge complex unit. Very limited extent of the alluvial tract, however, reduces the realisation of its productive potential.

The significance of fixed or stable dunes in retarding desertification needs no elaboration. These sandy hummocks are covered largely with *jal* (*Salvadora oleoides*), and to some extent with other forms of vegetation referred to previously in this chapter. It would be in the best interest of environmental management that these dunes are not reclaimed for agriculture or any other economic activity.

From the viewpoint of agricultural land-use and land-use efficiency, alluvium, marsh, waterlogged and potentially waterlogged land, and salt-affected land are more important cover types in the Sahibi alluvial tract. The alluvial soils are sandy to coarse loam in texture, and are referred to by Goyal and Singh (1982) as fluventic ustochrepts and typic ustorthents. These soils are low in fertility status, and suffer from periodic floods and subsequent erosion. Suitable soil conservation measures and application of recommended doses of farm-yard manures and fertilisers with normal agricultural practices are required to enhance the productivity of these soils. As the alluvial tract is endowed with good quality groundwater, a number of crops are raised using the subsurface water resources.

Marsh, waterlogged and potentially waterlogged, and salt-affected lands normally possess heavy textured saline-alkali soils referred to as typic natraqualfs (Goyal and Singh, 1982). Impeded drainage, low fertility soil status, low efficiency of fertiliser use, high fixation of nutrients, short tillage period, high watertable conditions, and presence of salts are some of the major problems encountered in these cover types. Suitable reclamation measures such as use of gypsum, provision of subsurface drainage, split dose application of fertilisers and farm-yard manure, and controlled irrigation should be applied to the salt-affected and potentially waterlogged lands. Salt tolerant crops like *dhaincha* and paddy are most suitable initially, and should be followed by paddy, *barseem* and barley. Mulching is also suggested for checking the upward movement of salts.

The structural hill environmental resource unit is subsumed into seven cover-types, of which hills, waterlogged and potentially waterlogged lands, salt-affected areas and pediments are the most widespread. The denuded Aravalli hill outliers are of no value to agriculture. Some sections of the hills have, however, been quarried for slate and limestone resources which provide employment to a section of rural population. Subdued hills, waterlogged and potentially waterlogged lands and salt-affected areas together blend into a geomorphic landscape of exceptional beauty in the northeast of the study area. These areas hold promise for the development of tourism; the Sultanpur bird sanctuary near Farruknagar is one such case in point.

Pediments, the product of parallel retreat of slope, are concave upward surfaces of gentle gradient that extend downslope from the hills. They are being described here for the first time from the semiarid environment of southern Haryana. All these pediments are subaerial surfaces as the basal part of there features lie buried beneath a cover of sand. A few hypotheses of pediments have been summarised by the author (Sharma, 1986a). Accordingly, pediments can, among others, originate from sheet floods, steam floods, stripping of regolith to the depth of solid rock, and rock-rotting due to localised concentration of moisture at the foot of hill mass. A few low-relief isolated short hills have also been observed within the confines of pediments. These may well be inselbergs, which aspect has not been examined in this research.

Pediment surfaces display an irregular cover of detritus overlying solid or partly decomposed parent rock. Therefore, these surfaces cannot be brought under plough. They constitute a part of the non-agricultural wasteland.

## Wasteland in Rewari Region

Despite the fact that farmers desire to cash on each and every conceivable parcel of land under the plough for the fear of government taking over uncultivated land for some other economic activity, a total of 407.68 out of 2887.75 sq km area of the region has been recognised in culturable and non-culturable wastelands (Table 3.5). Of this, 42.14 per cent is in culturable and the remaining 57.86 per cent in non-culturable wasteland category. Most of the culturable wasteland is in the undulating sandy plain environmental unit due largely to the growth of sodicity in soils — the after effects of recent developments in irrigation and to some extent due to rough terrain. The salt-affected lands can, however, be reclaimed by suitable treatment. An area of 4.38 sq km rendered waste due to brick-making activity can also be reclaimed for subsequent use. The entire area of 48.90 sq km in the fixed sand dune category is the other major component of culturable wasteland. The environmental propositions of the Rewari region do not, however, permit conversion of this cover-type to agricultural use. Further, as explained previously, the culturable wasteland in waterlogged terrain in the northeast of the study area (43.35 sq km) will have to stay as a permanent feature of the landscape, and cannot be reclaimed for agricultural use. Rocky protuberances in the undulating sandy plain, dune ridge complex, and Sahibi alluvium are a major component of non-culturable wasteland. Besides, an additional area of 160.46 sq km in hill outcrops, pediments, and broken land also constitute the bulk of non-culturable category of wasteland.

## Agricultural Land-use

Four villages situated respectively in the undulating sandy plain, dune ridge complex, salt-affected land, and Sahibi alluvium cover types were randomly selected to evaluate the productivity potential of these cover types in terms of the productive efficiency of crops raised in the above representative villages. The productive efficiency may be expressed by 'relative yield' and 'relative spread' indices.

TABLE 3.5: Classification of Wasteland in Rewari Region

(Area sq km)

| Environmental Resource Unit | Type/ Class* | Wasteland due to | | | | | | | | | | Total |
|---|---|---|---|---|---|---|---|---|---|---|---|---|
| | | Neglect/ excessive salts | Marsh | Depression | Water-logging | Sand dune | Abandoned brick kiln | Hill | Pediment | Rocky protuberance | Broken land | |
| Undulating sandy plain | 1 | 60.23 | — | 0.94 | — | — | 3.24 | | | | | 64.41 |
| | 2 | | | | | | | — | — | 68.78 | — | 68.78 |
| Dune ridge complex | 1 | 3.02 | — | 0.19 | — | — | — | | | | | 3.21 |
| | 2 | | | | | | | — | — | 4.28 | — | 4.28 |
| Fixed Dune | 1 | — | — | — | — | 48.90 | — | | | | | 48.90 |
| | 2 | | | | | | | — | — | — | — | — |
| Sahibi aluvium | 1 | 4.33 | 5.50 | — | — | — | 1.14 | | | | | 10.97 |
| | 2 | | | | | | | — | — | 2.35 | — | 2.35 |
| Structural hill | 1 | — | — | 0.97 | 43.35 | — | — | | | | | 44.32 |
| | 2 | | | | | | | 89.13 | 68.90 | — | 2.43 | 160.46 |
| Total | 1 | 67.58 | 5.50 | 2.10 | 43.35 | 48.90 | 4.38 | | | | | 171.81 |
| | 2 | | | | | | | 89.13 | 68.90 | 75.41 | 2.43 | 235.87 |

* 1. Culturable wasteland; 2. Non-culturable wasteland.

Whereas the relative yield refers to the yield of crop in the village to its average yield at the State level, the relative spread expresses the extent of crop coverage in the village to that of the district. Accordingly, the most efficient areas for a given crop cultivation are those where both the crops yield and its spread are high. Conversely, low yield-low spread crops represent areas of inefficient production. These two indices vary directly with the quality of land and the prevailing agro-environmental conditions.

The cropping pattern in 1978-79 of Tihara village in Mahendragarh district (S.O.I. topographical map No. $53^{D}/12$) has been taken to represent the productivity potential of the undulating sandy plain. A cropping intensity of 111.8 per cent reached in the village was supported by irrigation intensity of 21.6 per cent; the irrigation being from groundwater resources only. Nearly all of the *kharif* cropped area was in rainfed *bajra* (72.94%), fodder (14.21%) and *jowar* (11.84%). *Bajra* was determined to be a high yield-high spread crop, whereas the other two with low yield-high spread cover were noted to be low efficiency crops. These latter crops may be replaced by the better adapted *bajra* crop. Among the *rabi* crops, gram (42.33%), oilseeds (25.84%), barley (14.18%), and wheat (13.47%) accounted for nearly 96 per cent of the cropped area. Gram and oilseeds were grown under rainfed conditions, and wheat and barley were fully irrigated crops. Under rainfed conditions, gram is a low yield-high spread crop, while oilseeds represent a case of high yield-high spread crops. It is, therefore, possible to replace the area under rainfed gram to more remunerative oilseed crops. The productivity potential of undulating sandy plain land cover can be substantially enhanced by growing crops that are most suited to its geobiological environment which, among others, includes low inherent soil fertility, meagre rainfall, high evapotranspiratory demands, scanty surface water irrigation, and marginally saline to saline groundwater resources. Therefore, under rainfed conditions *bajra*-gram crop sequence may be replaced by bajra-oilseed crops and, in areas of limited irrigation wheat-barley and wheat-bajra will continue to be the two first ranking crops of the undulating sandy plain land cover.

Village Bhojawas in Mahendragarh district (S.O.I. topographical map No. $53^{D}/8$) has been taken to interpret the potential of the dune ridge complex. Limited irrigation from deep saline and marginally saline groundwater resources was determined in 1978-79 to be 17.2 per cent of the net area sown. For the same period, an exceptionally high cropping intensity of 179.11 per cent was also computed. A higher cropping intensity is probably a necessity to counteract frequent crop failures on a large scale. Thus, the area matured in the *kharif* and *rabi* was respectively 56 and 44 per cent of the cropped areas in the two cropping seasons. It is important to remember that enough volume of water of reasonable quality is not available for maintaining a cropping pattern of even moderate consumptive use. Therefore, choice in cultivating certain crops of a low water requirement is only limited. *Bajra* and fodder, the rainfed crops, respectively account for 48.8 and 6.9 per cent of the *kharif* matured area. *Bajra*, in particular, is a low yield-high spread crop in the geobiological environment of this land cover. Rainfed gram (28.4%) and oilseeds (5.56%), partly irrigated barley (5.07%), and fully irrigated wheat (4.28%) accounted for nearly all of the *rabi* matured area.

Except oilseeds which are high yield-low spread crops, the other three dominant *rabi* crops are normally low-yield and low to high spread crops. The productive efficiency of all the crops is, therefore, low in this land cover of very poor productivity potential. Irrigated gram, however, provides high yields per unit area, and has a high spread. It is also observed that irrigation does not make any quantitative change in the yield of barley when compared to its yield under rainfed conditions. It may, therefore, be suggested that priorities in irrigation be shifted from barley to gram for enhancing the productivity potential of this land cover.

The cropping pattern in Chhapar village of Rohtak district (S.O.I. topographical map No. $53^{D}/11$) revolves around rainfed and minor irrigation conditions in a land cover that has been photo-interpreted to be effected slightly by salt-hazards. Rainfed *bajra* (65.76%) and *jowar* as fodder (21.28%) are respectively high yield-high spread and high yield-low spread crops. Fully irrigated sugarcane (4.53%) is a perennial crop. Gram, irrigated in 48 per cent of its cropped area, is the first-ranking crop spread over 39.5 per cent of the *rabi*-season crop land. It is followed by wheat and oilseed crops that respectively account for 31.2 and 16.9 per cent of the total cropped area. Of these, gram is a high yielding crop under both rainfed and irrigated conditions. Whether irrigated or not, the wheat crop in the salt-affected cover type remains a low yield-high spread inefficient crop. Similarly, oilseed crops are also determined to be low-yield crops under irrigated conditions. The perennial sugarcane is rated to be high yield crop, and it can easily replace the inefficient crops like wheat and oilseeds. Besides, sugarcane is also adapted to a high soil-salt environment. As irrigation facilities will expand, and these may have expanded since 1979, the salt hazards are likely to become more pronounced requiring adoption of such crops as are salt-tolerant.

A diversified cropping pattern in village Reliawas of Mahendragarh district (S.O.I. topographical map No. $53^{D}/12$) represents the ecological environment of the Sahibi alluvium land cover. The cropping pattern has developed under a moderate irrigation intensity of 53.1 per cent and under conditions of rainfed agriculture, which together give a cropping intensity of 164.6 per cent in the village. Despite better developed minor irrigation facilities, the *bajra*-wheat-gram cropping pattern reflects the inadequacy of irrigation to support crops with higher consumptive use. It may be noted that bajra and gram are rainfed and wheat a partly irrigated and partly rainfed crop. *Kharif* fodder with some 20 per cent of the area under its cultivation is irrigated. The cumulative area under other irrigated *kharif* crops like pulses, rice and sugarcane is, however, less than 3 per cent. In this environment, *bajra* is a high-yielding crop, but *jowar* used mostly as a fodder is an inefficient crop with low yields per unit area. The rainfed gram (25.9%) is a high yield-low spread crop, while wheat-gram mixture grown under similar practices is a low yield-low spread inefficient crop. Some 60 per cent of wheat hectarage of 39.4 per cent of *rabi* cropped area is irrigated. Irrigated wheat provides high yields, but it is a low yield crop under rainfed conditions. Rainfed barley raised on residual soil moisture gives high yields. Similar yields are also reported when it is irrigated. Rape and mustard are high yield-low spread oilseed crops when raised with irrigation. From the above discussion, a cropping pattern most suitable to the Sahibi alluvium land cover may be suggested as : Rainfed *bajra*-irrigated wheat-rainfed gram-rainfed barley-irrigated oilseeds. Thus, discriminant use of irrigation water can enhance the productivity potential of the Sahibi alluvial cover type.

# 4

# LAND RESOURCES AND LAND RESOURCE PLANNING IN NUH REGION

## Introduction

COVERING 592.87 sq km, the Nuh area extends approximately from latitudes 27°59'55" to 28°12'48" north, and longitudes 76°53'20" to 77°15' east. Nuh and Malab in Gurgaon district, and Hattin in Faridabad district are some of the important settlements of the region.

Low incidence of mean annual rainfall, brackish subsurface water, and non-availability of canal irrigation in the past had led to periodic famines in this area of low productivity agriculture. With the commissioning of the flow-cum-lift irrigation Gurgaon Canal scheme, the productivity level of the land has, however, increased in the recent times. Except the northwestern section and the Aravalli hills in the west, a part of the area is now irrigated by the Gurgaon Canal and its distributary network by free-flow and lift techniques.

The Nuh area experiences 85 rainy-days during which time 58 cm of rainfall is received. This rainfall has an annual variability of 30 per cent. The seasonal variability of rainfall is computed at 45 per cent and 90 per cent respectively for the summer and winter seasons (Rani, 1978). In this area of high thermal efficiency, in which the maximum temperatures throughout the year vary between 20°C and 46°C, the evaporation is high. The annual evaporation at Delhi, a few kilometres to the north of the area, is determined to be 203 cm (Goyal *et al.*, 1978). Given the above conditions, the Nuh region is typical of a semiarid climate.

## Topography and Geology

The region is topographically segmented broadly into two non-contiguous plains and the Aravalli hills, some of which have been considerably worn-down to a low relief. The plain between Nuh and Hattin is nearly a flat surface with few depressions, in which the elevation varies from 190 to 207 m. A few Aravalli hill outliers, in elevation between 209 and 250 m, also traverse through this area. Several remnants of palaeochannels representing either inland drainage from the bounding Aravalli hill range in the west, or a through southward-directed drainage, are also recognised on aerial photographs. This plain also is typical of wetland terrain as is evident from 20-30 per cent of the area shown in swamps on 1910-11 Survey of India topographical map. These wetlands have since been largely reclaimed. The plain to the northwest of the region has a relief of 270 to 285 m, and represents a broad rolling character. Several short-segmented seasonal torrents drain north and northwestward through the plain. The main, north trending Aravalli hill ridge rises to a maximum elevation of 405 m above and the mean sea level in the study region.

The Nuh area comprises of the sediments of three distinct provinces divisible into two major geological periods. These are aeolian and still-water deposits of the Recent, and the Aravalli hills and their few outcrops of the Precambrian tectonic origin. The Aravalli hill range is a relic of a once mighty Aravalli System that has been subject to varying degrees of metamorphism and folding. These hills are composed of Delhi quartzites, slates and phyllites. The phyllites occupy a lower stratigraphic position in the system, and are surrounded by sediments of later periods. The Recent sediments of aeolian and still-water origins have developed vast plains with their own characteristic environmental hazards, and limitations for land-use.

## Geomorphic Environments

The Precambrian Aravallis of tectonic origin, and the Recent earth-surface environments of inland lake evaporite, and dryland are represented in the Nuh region of extreme topographic diversity (Fig. 4.1). The morphology of Aravalli hills is an embodiment of a long period of wear and tear of rocks that were subject to climatic changes of the past. These morphological attributes are probably in equilibrium with the present-day climatic environment. The areal extent of environmental resource units is given in Table 4.1.

TABLE 4.1: Area in Different Environmental Units in Nuh Region

| Geomorphic environment | Period | Environmental Unit | Area (sq km) |
|---|---|---|---|
| Dryland | Recent | Alluvial apron | 0.67 |
| Environment | Recent | Sandy plain | 26.55 |
| | Recent | Broken land | 2.17 |
| | Precambrian | Structural hills | 40.25 |
| Inland Lake | Recent | Lake flat | 521.43 |
| Evaporite Environment | Precambrian | Structural hills | 1.80 |
| Total | | | 592.87 |

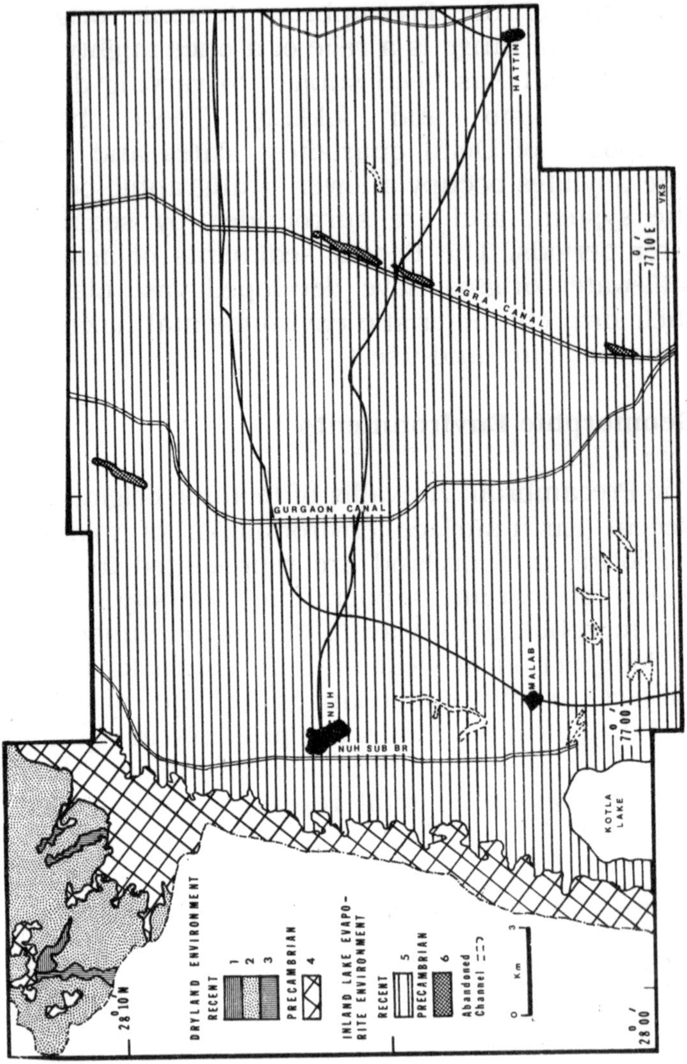


FIG. 4.1: Classification of environment and associated landforms in Nuh region. Dryland Environment: *Recent* - 1. Alluvium, 2. Sandy plain, 3. Broken land; *Precambrian* - 4. Structural hills. Inland Lake Evaporite Environment: *Recent* - 5. Lake flat; Precambrian - 6. Structural hills.

The alluvial tract from near Nuh and extending eastward to Hattin and beyond is a part of the Firozepur Jhirka valley. This longitudinal syncline is bounded in the west and east respectively by the Sohna-Owganwa and Mandkaula-Biwan Aravalli hill ranges. According to Awasthi, Prashad and Chatterjee (1976), the plain grew out of an inland Nuh lake. This large inland depression is now occupied by the remnants of Kotla and Ujina lakes. The above authors suggest that Recent inland drainage from north and south draining into the depression that extended down to the bedrock gradually accumulated fine-grained sediments in the still-water environment of the lake. The presence of grey-coloured sticky clays and absence of sand in boreholes, particularly at Malab, Salheri, and Adlaunka (see, Table 4.2) suggests these to be the deepest part of the lake. Further, calchie mixed with clays and the presence of gypsum in some of the boreholes suggests that these are evaporite deposits. The evaporite environment is further supported by the presence of common salt and salt-peter, the extraction of which formed the basis of mining industry in this area until the very recent time.

The dryland environment in this region is represented by a widespread undulating cover of aeolian sand of variable thickness, and an alluvial apron of limited extent. That the inland drainage normally makes deep and narrow channels over the sandy plain probably suggests the presence of rockbed at shallow depth beneath the sand cover. It is also possible that channel development, that may have initially occurred over the sloping piedmont/pediment of the Aravalli range, predates desertification leading to expansion of sandy accumulations in this area.

TABLE 4.2: Borehole Data from Selected Sites
Village - Nasirbas (27°51'N, 76°58'E)

| Lithology | Depth range (m) | Thickness (m) |
|---|---|---|
| 1 | 2 | 3 |
| Top soil, sand and clay | 0-0.92 | 0.92 |
| Clay, brownish, pellets of greenish clay, slight sand, with a few pieces of calcareous nodules | 0.92-7.32 | 6.40 |
| Clay, brownish, with soft and hard bands, plastic, variable amounts of calcareous nodules | 7.32-111.25 | 103.93 |
| Clay, mottled, with soft and hard bands slightly sandy, plastic, with variable amounts of calcareous nodules | 111.25-148.74 | 37.49 |
| Quartzites and slates in alternative bands. Quartzites are fine-grained, bluish and slightly greenish | 148.74-153.75 | 5.01 |
| Village - *Malab* (28°01'N, 77°01'E) | | |
| Clay, greyish brown, and *kankar* with little silt | 0 -6.10 | 6.10 |
| Clay, greyish brown, sticky, with little *kankar* | 6.10-10.67 | 4.57 |
| *Kankar* with clay | 10.67-13.72 | 3.05 |
| Clay, greyish brown,sticky with very little *kankar* | 13.72-20.73 | 7.01 |
| Clay, brownish yellow, sticky with very little *kankar* | 20.73-45.72 | 24.99 |
| Clay, greyish brown, sticky | 45.72-62.79 | 17.07 |
| Clay, greyish brown, sticky, with very little *kankar* | 62.79-122.83 | 60.04 |
| Clay, greyish brown, sticky with *kankar* | 122.83-160.93 | 38.10 |
| Clay, greyish brown, hard and sticky, with *kankar* | 160.83-199.94 | 39.01 |
| Weathered crystalline rock with little clay | 199.94-201.46 | 1.52 |
| Black crystalline limestone | 201.46-201.92 | 0.46 |
| Village - *Salaheri* (28°06'N, 77°02'E) | | |
| Clay, greyish brown, sticky, with little *kankar* | 0 -6.09 | 6.09 |
| Clay, greyish brown, sticky, with very little *kankar* | 6.09-69.80 | 63.71 |
| Clay, greyish brown, sticky, with little *kankar* | 69.80-174.95 | 105.15 |
| Clay, greyish brown, very hard, sticky, with *kankar* | 174.95-206.96 | 32.01 |

| | | |
|---|---|---|
| Clay, reddish brown, sticky, with little *kankar* | 206.96-220.96 | 14.00 |
| Weathered hard crystalline rock | 220.96-221.73 | 0.77 |
| Black crystalline limestone | 221.73-222.20 | 0.47 |
| Village - *Aldaunka* (28°07'N, 77°07'E) | | |
| Clay, greyish brown, with *kankar* and silt | 0 -3.05 | 3.05 |
| Clay, greyish brown, sticky, with little *kankar* and silt | 3.05-10.67 | 7.62 |
| Clay, greyish brown, sticky, with very little *kankar* | 10.67-45.72 | 35.05 |
| Clay, brownish yellow, sticky, with very little *kankar* | 45.72-104.85 | 59.18 |
| Clay, greyish brown, sticky, with very little *kankar* | 104.85-146.91 | 42.06 |
| Clay, greyish brown, very hard and sticky, with very little *kankar* | 146.91-291.03 | 84.12 |
| Weathered, black crystalline limestone with clay | 291.03-298.04 | 7.01 |
| Village - *Bhadas* (27°55'30" N, 77°01'E) | | |
| Top soil, sand and clay | 0 -0.92 | 0.92 |
| Clay, light brown, sandy | 0.92-3.96 | 3.04 |
| Clay, black, plastic, occasionally sandy, with calcareous nodules and shells | 3.96-14.33 | 10.37 |
| Clay, light brown and occasionally black, plastic, with variable amounts of calcareous nodules and shells (towards top) | 14.33-52.12 | 37.79 |
| Clay, light brown, plastic, with variable amounts of calcareous nodules and a very few gypsum pieces | 52.12-67.05 | 14.93 |
| Clay, light brown, plastic, with calcareous nodules | 67.05-71.93 | 4.88 |
| Clay, light brown, plastic, with variable amounts of calcareous nodules and occasional gypsum | 71.93-157.28 | 85.35 |
| Slate, black | 157.28-164.28 | 7.00 |

*Source:* Geological Survey of India, *Exploratory Tubewell Organisation* (Abstracted from S.C. Awasthi, Mangla Prashad and A. Chatterjee, 1976).

Alluvial apron deposited by some of the larger rivulets has also been recognised on aerial photographs. The deposition of alluvium is related to seasonal transport of weathered products from the source area of these seasonal streams in the Aravalli hills.

## Discussion

Land resources in the Nuh area of contrasting landscape may individually be examined and assessed in respect of their geomorphic environment. The potential and limitation of the land resource base for exploitation and optimum utilisation has, thus, been studied separately for the two major geomorphic environments discussed in the previous section. The Aravalli hills range has been treated as a part of the dryland environment as it possesses morphological attributes typical of the above-said environment. However, the low-relief outliers of the Aravallis within the inland lake evaporite environment have been treated as a component of the environment in which they are dispersed.

## Resource Units of the Dryland Environment

The dryland environment has been recognised in 69.64 sq km of the area (Table 4.1). The land cover given in Fig 4.2. and quantitatively analysed in Table 4.3 shows the spatial pattern of various land resource units of this environment.

Segmented by broken land, rocky protuberances, alluvium, and salt-affected soils, the sandy plain is an extensively cultivated land resource unit. It has ustipsamment soils with predominantly sandy to sandy loam soils. Sandy soil texture, seasonal shifting of loose sand, non-availability of canal irrigation, and prevalence of marginal to saline subsurface water place serious constraints on the productive use of this resource unit.

TABLE 4.3: Area in Different Cover Types, Including Built-up Land in the Dryland Environment of Nuh Region

| Cover type | Area (sq km) | Percentage of total area |
|---|---|---|
| Sandy plain | 22.93 | 32.93 |
| Hill | 32.71 | 46.97 |
| Broken land | 4.96 | 7.12 |
| Rocky protuberance | 3.90 | 5.60 |
| Salt-affected land | 2.15 | 3.09 |
| Waterlogged and potentially waterlogged land | 1.47 | 2.11 |
| Alluvium | 0.67 | 0.96 |
| Orchard/grove | 0.05 | 0.07 |
| Wasteland, including village common land | 0.80 | 1.15 |
| Total | 69.64 | 100.00 |

According to the studies reviewed by Williams (1985), arid conditions in India were accentuated around 4000 years BP. This late-Holocene desiccation probably reactivated the dune-building activity in the Thar and, consequently, advanced the shifting sand further east to built the sandy plain.

The Aravalli hills alone accounted for 46.97 per cent of the 69.64 sq km area in the dryland environment. These denuded hills are agriculturally or otherwise economically unproductive land resource unit.

Broken land in the form of linear ribbons is recognised along seasonal torrents draining the north-facing Aravalli slopes. These gullyed lands are 2 to 7 m deep from the surrounding area. They retain their depth even as the gullys traverse through the sandy plain. Such a cross-sectional

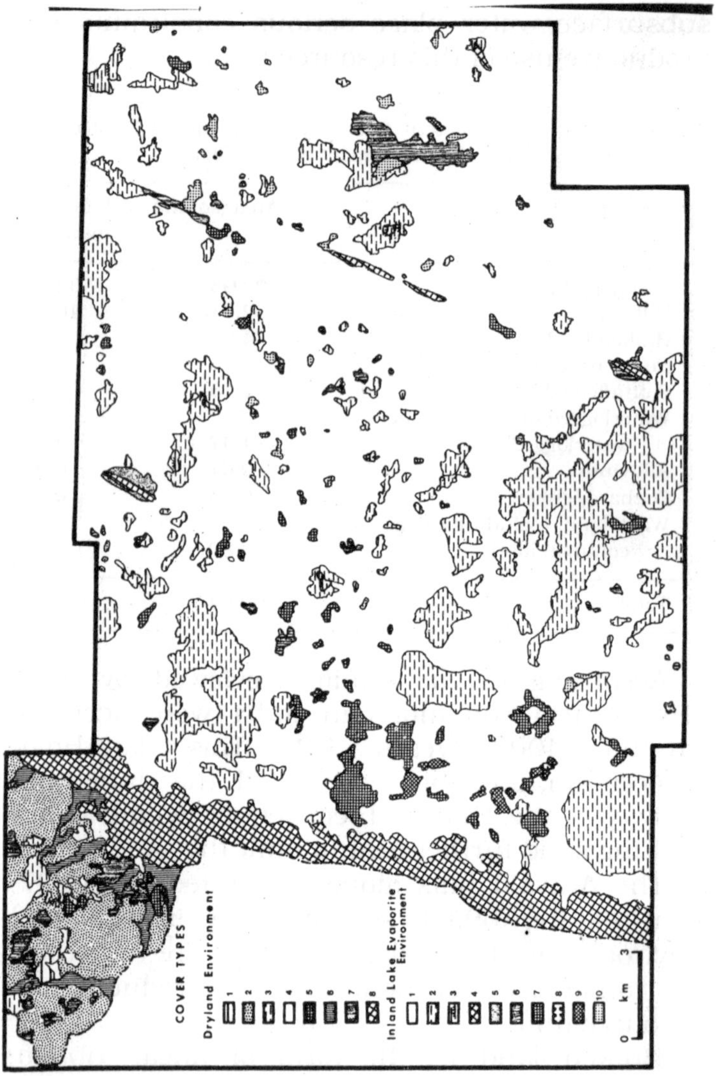


FIG. 4.2: Cover types in Nuh region. *Dryland Environment*: 1. Alluvium, 2. Sandy plain, 3. Waterlogged and potentially waterlogged land, including water bodies, 4. Salt-affected land, 5. Wasteland, 6. Broken land, 7. Rocky protuberance, 8. Aravalli hills. *Inland Lake Evaporite Environment*: 1. Lake flat, 2. Waterlogged land, including water bodies, 3. Marsh, 4. Aravalli hills, 5.' Pediment, 6. Rocky protuberance, 7. Wasteland, 8. Fixed sand dune (area less than 0.01 sq km), 9. Brick kiln, 10. Grazing land.

character probably suggests presence of either alluvium or the Aravalli rocks at shallow depth beneath the sandy cover. It is also possible that their development predates the deposition of sandy plain. This unit is also unproductive in terms of utility and economic returns.

Rocky protuberances, that may not always be visible on the surface, are recognised on the tonal and textural contrasts on airphotos. They represent the flattened, buried or exposed, parts of the receding and intensively weathered Aravalli hills. This land resource unit is typic ustorthents, and presents difficulty of tillage.

The salt-affected lands generally occur in association with high watertable and seasonal wetting and drying of surficial materials along stream courses and foothills with gentle slopes. These lands are characteristic of sharp variations in the physico-chemical properties of sediments that vary in origin from wind, and sheet and stream transported to solum on hill slopes.

Waterlogged and potentially waterlogged land cover only 1.47 sq km of the 69.64 sq km surveyed area. One such occurrence is observed south of Guarka and north of Buraka villages where seasonal runoff from one of the numerous seasonal torrents is blocked by an embankment, creating seasonal storage of surface runoff. This water of nearly fresh quality can be used for irrigation during the *kharif* season. The storage also augments the groundwater reservoir locally. Further, the land reclaimed after the ponded water has been used-up is tilled for raising at least one crop in the *rabi* season. Such a practice is good illustration of water and soil conservation techniques, and must be adopted for other areas with similar geomorphic characteristics.

Despite numerous streams debouching from the Aravalli slopes, the occurrence of alluvial deposits is rather limited on account of deep and narrow cross-sectional characteristics of the stream channels. Only where the level of stream beds approaches that of the surrounding area and the thalweg flattens that alluvium is deposited as overbank sediments during storm-period runoff. This land resource unit is favoured with loamy soils, flat surface, and shallow fresh water resources. The limited extent of this unit may, however, impose restrictions on its optimum utilisation.

### Wasteland in the Dryland Environment

As shown in Table 4.4, all the wastelands covering 59.95 per cent of the surveyed area are in the non-culturable category. Hence, these lands in Aravalli hills, broken land, and rocky protuberance are not available for agricultural or any other economic use. The absence of culturable wasteland may be explained by the fact that farmers have to make the best use of all the available land for agricultural purpose in this area of low-productivity environment.

### Resource Units of the Inland Lake Evaporite Environment

The two major landscape units within the generally clayey still-water sediments covering 523.23 sq km area (Fig 4.1) are the remnants of Aravalli hills and their associated features, and the alluvial plain with its characteristics land cover associations. The land cover of the inland lake evaporite environment has been mapped in Fig 4.2, and its quantitative aspects detailed in Table 4.5.

TABLE 4.4: Classification of Wasteland in the Dryland Environment of Nuh Region

| Type/class | Area (sq km) | | | | | | |
|---|---|---|---|---|---|---|---|
| | Neglect/ excess-ive salts | Aban-doned brick kiln | Hill | Broken land | Rocky prot-uber-ance | Total | Percent-age of surveyed |
| Culturable wasteland | – | – | – | – | – | – | – |
| Non-culturable wasteland | – | – | 32.71 | 4.96 | 3.90 | 41.57 | 59.95 |
| Total | | | | | | 41.57 | 59.95 |

## Land Cover

Only a few buried low-relief remnants of Aravalli hills covering 1.80 sq km are observed to the east of Agra Canal and near the village Mailawas. These hills either rise sharply from the surrounding sedimentary mass or form distinct piedmont. The piedmont covers only 0.14 per cent of the surveyed area. Due to constraints of hard rock surface and slope, these land resources units do not offer opportunities for agricultural development. The rocky protuberances covering 6.71 sq km area or 1.28 per cent of the area analysed probably represent the low-relief exposed and buried rocks at shallow depth. This land resource is also not known to support agricultural activity.

Salt-affected land, waterlogged and potentially waterlogged land, water bodies, and marsh-land account for 95 per cent of the surveyed area in the inland lake evaporite environment. Varying degrees of grey stone suggesting low to very severe salt-hazards in what appears on aerial photographs to be relatively compact surficial material is classified as salt-affected land. As elsewhere in this research, the term salt-affected land denotes both the alkali and saline soils. The large extent of salt-affected lands observed in the Nuh region is in conformity with the litholog data given in Table 4.1. Parts of this terrain are classified by Goyal *et al.* (1978) in fine to coarse loamy saline-alkali, and calcareous and non-calcareous soils belonging to typic ustorthents and typic camborthids soil taxonomic classes. Further, the application of brackish groundwater for irrigation has also enhanced the saline-alkali hazards in these level to nearly level lands that are ideal for extensive cultivation. The April 1986 CAD Project Report for the Gurgaon Canal Command notes that the EC of groundwater is as high as 30,000 micromhos/cm, and in some areas the residual sodium carbonate is observed to be in excess of 14 Me/litre.

TABLE 4.5: Area in Different Cover Types, Including Built-up Land in the Inland Lake Evaporite Environment of Nuh Region

| Cover type | Area (sq km) | Percentage of total area |
|---|---|---|
| Hill | 1.80 | 0.34 |
| Piedmont | 0.73 | 0.14 |
| Rocky protuberance | 6.71 | 1.28 |
| Salt-affected land | 427.41 | 81.69 |
| Waterlogged and potentially waterlogged land, and water bodies | 70.17 | 13.42 |
| Marsh | 2.05 | 0.39 |
| Orchard/grove | 0.18 | 0.03 |
| Brick Kiln | 0.88 | 0.17 |
| Grazing land | 3.32 | 0.63 |
| Wasteland, including village common land | 9.98 | 1.91 |
| Total | 523.23 | 100.00 |

The waterlogged and potentially waterlogged lands, water bodies, and marshes covering 13.81 per cent of the 523.23 sq km surveyed area are related directly with the still-water accumulation of sediments in the inland lake. This is to emphasise here that the wetlands are typical association of this depositional environment. The waterlogged and potentially waterlogged conditions are a consequence of shallow watertable that locally is only 0.56 m from the surface (CAD, April 1986), relatively impervious nature of sediments (Table 4.2), and seepage from canal irrigation network.

Kotla lake is the most conspicuous water body among several natural depressions widely dispersed in the Nuh region. The lake collects surface runoff during the rainy season from a large inland catchment area within the Ferozepur Jhirka Valley. Consequently, water level in the lake fluctuates from one rainy season to the other. The Kotla lake may be regarded as a playa lake.

Extensive marsh-lands that had existed in the early 1920s have, as noted previously, now been largely reclaimed for agricultural purposes. The effects of reclamation of these wet-lands on ecology is, however, not known in definite terms. According to the study by Goyal and others (1978), the areas classified here as waterlogged and potentially waterlogged lands, and marshes are clayey, saline and saline-alkali, non-calcareous to calcareous hyperthermic, vertic camborthids and aquicvertic camborthid soils.

The other land covers recognised in the Nuh region are orchard/grove, brick kiln, grazing land, and wasteland spread over 14.36 sq km area. As noted in Table 4.6, the extent of culturable wastelands is marginally higher than the land in non-culturable wasteland category. The culturable wasteland comprises of lands made unproductive due to intense saline-alkali hazards, marshes and abandoned brick kilns. The bulk of non-culturable wasteland is due to rocky proturberances.

## Agricultural Land-use

The Command Area Development Authority (April 1986) has mapped the land capability and examined the related crops for a major part of the Nuh region. This aspect, briefly summarised here, portrays that the geomorphic criterion for terrain evaluation also very nearly explains the land capabilities of the inland lake evaporite environment region. The agricultural productivity of this area have been further referred to the crops raised at Malab (S.O.I. toposheet No. 53 H/4).

TABLE 4.6: Classification of Wasteland in the Inland Lake Evaporite Environment of Nuh Region

| Type/Class | Area (sq km) | | | | | | | |
|---|---|---|---|---|---|---|---|---|
| | Neglect/excessive salts | Marsh | Abandoned brick kiln | Hill | Piedmont | Rocky protuberance | Total | Percentage surveyed area |
| Culturable wasteland | 6.91 | 2.05 | 3.07 | – | – | – | 12.03 | 2.30 |
| Non-culturable wasteland | – | – | – | 1.80 | 0.73 | 6.71 | 9.24 | 1.76 |
| Total | | | | | | | 21.27 | 4.06 |

According to the study referred to above, most of the region is shown to possess Class III capability, with the remaining area in non-contiguous Class II, Class IV, and Class VIII lands. Class II land possesses fertile soils which, however, suffer from alkali hazards. Here, rainfed *bajra, jowar*, barley and *dhaincha*, and wheat, sugarcane and vegetables under irrigated conditions are raised. Class II land is said to be ideal for growing additional crops such as maize and pulses under rainfed conditions.

Class III land suffers from a host of inherent limitations in its various subclasses that refer to medium to coarse-grained soils, poor to moderate moisture holding capacity of soils, occasional stagnation of water during rainy season, poor surface drainage, and moderate salinity-alkalinity problems. Consequently, each subclass requires to be developed for specific crops and crop combinations. In general, double cropping with irrigation is recommended for these lands.

Class VIII capability land refers to eroded Aravalli hills, piedmont and rocky protuberances. These lands are unsuitable for commercial plant production, but should be developed for watershed management, and for aesthetic and recreational purposes.

The cropping pattern of Malab in 1978-79 showed the dominance of unirrigated *jowar, bajra* and barley in the *kharif* season, and wheat, gram and wheat-gram mixture in the *rabi* season under irrigated conditions. All of this cultivated land is shown to be in Class III capability class. It appears, therefore, that the above crops are in adjustment with the physical environment. The productivity level of this area may, however, be enhanced by adopting suitable land, water, and crop management practices.

# 5

# LAND RESOURCES AND LAND RESOURCE PLANNING IN FIROZEPUR JHIRKA REGION

## Introduction

THE FIROZEPUR Jhirka Region is spread over 495.56 sq km area in parts of Gurgaon and Faridabad districts in south-eastern Haryana. The region extends between 27°45' and 27°55' north latitudes and, 76°54'30" and 77°44'20" east longitudes. Firozepur Jhirka, Punhana, Nai, Bichhor and Hodal from west to east are some of the major settlements of this area.

The region represents three diverse geological environments: inland lake evaporite environment, dryland environment, and fluvial environment (Fig. 5.1). As noted in previous chapters, each of these environments and sub-environments present diverse geomorphic landscapes comprising of typical landforms which offer specific potential and limitation for the use of land resources. Following the consistent methodological treatment of the subject content of this research, the landforms of these specified environments have been independently studied and evaluated for their overall performance *vis-a-vis* sustained economic growth in context of rural development. The areal extent of environmental units is given in Table 5.1.

The average rainfall recorded over various raingauge stations in Gurgaon and Faridabad districts is reported to be 46.3 and 55.8 cm respectively. Nearly 58 per cent of this rainfall amount is registered during July and August in the monsoon season. About 2 cm of rain from westerly disturbances is also received from December through February months. The temporal pattern of rainfall is such that long periods of dry-spells, and sometimes absolute drought conditions commonly prevail in the region. These long dry-spells are critical to the development of rainfed agriculture, and are of special significance to irrigated agricultural practices where surface water availabilities are erratic and subsurface water resources generally brackish. The temperature recorded at Gurgaon shows that the maximum temperature rises to 46°C in June and the minimum temperature occasionally drops to 0°C in the month of January. High temperatures throughout the year and paucity of meteoric moisture provide the region a characteristic semiarid climate.

TABLE 5.1: Area in Different Environmental Units in Firozepur Jhirka Region

| Geomorphic environment | Period | Environmental unit | Area (sq km) |
|---|---|---|---|
| Fluvial Environment | Recent | Meander-cut Plain | 31.68 |
| | Recent | Dry Valley | 16.96 |
| | Pleistocene | Alluvial Upland | 253.76 |
| | Precambrian | Structural Hills | 1.07 |
| Dryland Environment | Recent | Alluvium | 1.01 |
| | Recent | Sandy Plain | 41.92 |
| | Recent | Fixed Sand Dune | 0.83 |
| | Precambrian | Structural Hills | 99.37 |
| Inland lake | Recent | Lake Flat | 48.12 |
| Evaporite Environment | Precambrian | Structural Hills | 0.84 |
| Total | | | 495.56 |

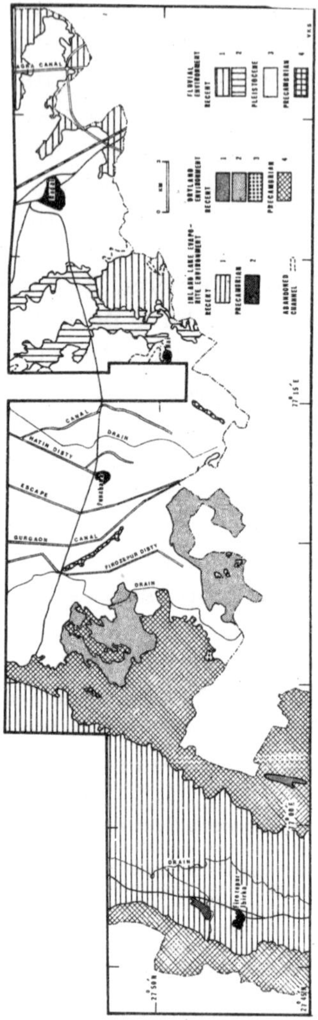

FIG. 5.1: Classification of environment and associated landforms in Firozepur Jhirka region. Inland Lake Evaporite Environment: *Recent* - 1. Lake flat; Precambrian 2. Structural hills. Dryland Environment: *Recent* - 1. Alluvium, 2. Sandy plain, 3. Fixed sand dune; *Precambrian* - 4. Structural hills. Fluvial Environment: *Recent* - 1. Meander-cut plain, 2. Dry valley; *Pleistocene* - 3. Alluvial upland; *Precambrian* - 4. Structural hills.

## Inland Lake Evaporite Environment

The inland lake evaporite environment in Firozepur Jhirka region is identical in respect of form process and material composition to that already discussed for the Nuh region in Chapter 4. Further, the two regions with similar climatic regimes should logically be also similar in inherent potential and productive level of land resources, and offer parallel set of limitations for their development. A few differences in regional development of the two regions are, however, possible on account of variations in anthropogenic impacts on land resources.

## Environmental Resource Units

The inland lake evaporite environment in 48.96 sq km or 9.88 per cent of the area analysed in this region comprises of the Recent lake flat and Precambrian structural hills, of which the former alone covers 98.28 per cent of the total area; the remaining being in the Aravalli hills. The elevation of the fairly level lake flat gradually declines from 214 m in the southwest to 191 m in the northeast. A few micro-order relief variations in the form of isolated broad depressions are, however, common in this resource unit. Within this geomorphic environment, two structural outliers of the Precambrian Aravalli hills steeply rise to heights of 40 to 100 metres from the surrounding sedimentary mass, and respectively reach 215 and 318 metres above the mean sea level.

## Dryland Environment

The dryland environment in the Firozepur Jhirka region is largely dominated by the Precambrian structural and lithologic fabric of the Aravalli hills, fairly extensive Recent sandy accumulations in the form of sandy plain and fixed sand dunes, and alluvial cover of limited extent. The dryland environment has been identified in 143.13 sq km area of the region. The geological characteristics of this environment have been described previously for the Rewari region (Ch. 3); the differences being mainly in the intensity and magnitude of geomorphic processes that presently operate or have operated in the past. The geomorphic processes presently operating in the Firozepur Jhirka region are little less vigorous than those obtaining in the Rewari region. Four environmental units as alluvium, sandy plain, fixed sand dune, all of the Recent, and Precambrian structural component of the Aravalli System have been recognised in the area under reference (Fig. 5.1 and Table 5.1).

## Environmental Resource Units

A small stretch of alluvial cover (1.01 sq km) is recognised in the bowl created in the gap through the Aravalli hills, and the receding hill segment into which seasonal surface flow converges. Apparently, the alluvium is thin, and overlies the crystalline substratum.

The sandy plain is an undulating surface of a variable thickness of fine to medium sand overlying the rockbed. Two dispersed occurrences of the unit have been identified in the region. The unit in the north is a southward sloping surface in which the elevation varies between 193 and 200 m. The unit in south varies in elevation from 188 to 192 m and has a westward gradient. A few patches of fixed dunes have also been observed in this occurrence of the sandy plain. The sandy plain resource unit is one of low productivity due to limitations of shifting sands, grain size composition, and low moisture holding capacity of the sands. On the basis of topographic associations, the sandy plain is reasoned to have developed from the passage of sand-laden wind across the hills and through gaps in the Aravalli outliers.

Fixed dunes are associated with the lee of structural hills and sandy plain. In the former case, the fixed dunes are probably formed as sand shadows in the zone of dead-sand. The deposition is temporarily caused by cleavage of the principal wind current as it passes over the hill ranges. In the latter case, they rise above the sandy plain as low relief sand mounds stabilised by the growth of *sarkanda* (*Saccharum munia*), *ak* (*Calotropis* spp.), and several other shrubs, bushes, weeds and grasses. These dunes may also be remnants of larger dunes, parts of which have been flattened for very low productivity dryland farming.

Enclosing the inland lake evaporite environment, the structural hills cover 99.37 sq km or 69.42 per cent of the dryland environment. This unit includes all areas of Precambrian geological structure and lithology from hill ranges to rocks worn-down to the level of the adjoining Pleistocene upland surface. The elevation of hill ranges, of isolated hills and that of the other geomorphic features of the unit declines towards the north and northeast. Thus, the western and eastern sections of the ridges respectively decline in elevation from 437 to 385 m, and 356 to 242 m. Likewise, the individual hill outcrops vary in elevation from 311 to 242 m, and the intensely weathered and levelled-down sections are from 211 to 216 m in elevation.

These denuded hills of metamorphic and

crystalline rocks are of no consequence to agriculture, and require afforestation and watershed management practices to check the menace of seasonal erosion of land. The hills, however, can offer adequately suitable environment for the development of tourism.

## Fluvial Environment

The fluvial environment, recognised in 303.47 sq km area, encompasses the Recent, Pleistocene, and Precambrian geological periods (Table 5.1). Meander-cut plain and dry valley respectively covering 31.68 and 16.96 sq km areas are the principal environmental resource units of the Recent. The Pleistocene alluvial upland (253.76 sq km) and Precambrian structural Aravalli hills (1.07 sq km) are the other two environmental units within the framework of fluvial environment.

The fluvial environment in the Firozepur Jhirka region is only a southward extension of the geological processes discussed for the Sonipat region (Ch. 3). Hence, the geological aspects of alluvium have not been repeated for the reference region under discussion.

## Environmental Resource Units

The meander-cut plain is the product of eastward migrating Jamuna river during the recent geological period, and is identical in morphology with the one described for the Sonipat region. The meander-cut plain varies in elevation from about 185 to 190 m above the mean sea level. A few abandoned channel depressions observed in this unit are photo-interpreted to the rockbed channels. This observation indicates a shallow nature of the alluvial cover over the crystalline basement rocks.

The resource unit is generally free of salt hazards. However, irregular sharp surface slope variations, caused by wide swings of the meander-belt of the Jamuna river, may offer difficulty in extensive use and efficient management of this land resource. The meander-cut plain is also placed favourably in respect of generally good quality groundwater at shallow depth. This aspect is of immense significance to the development of and higher agricultural productivity per unit area from this resource unit.

The dry valley is an erosional fluvial feature situated higher in elevation than the meander-cut plain. It has developed entirely up on the upland alluvial tract as a linear feature of shallow depth, but up to 2.4 km wide in cross-section. The forked nature of the valley indicates it to be a likely conduit for southward-directed surface drainage during the rainy season, and that it is not an abandoned channel of the Jamuna river.

The unit is normally free of salt-hazards. Here the surface drainage is also exceptionally good. Further, good quality subsurface water resources occur at shallow the medium depth. All these parameters enable the land resources of the dry valley to be intensively developed for higher productivity agriculture.

The Pleistocene alluvial upland tract is the most extensive environmental resource unit that covers 253.76 sq km area in the Firozepur Jhirka region. The tract is situated 5 m or more above the adjacent meander-cut plain, and varies from about 190 to 196 m in elevation above the mean sea level. A few abandoned channels, interpreted as rockbed channels, occur at the contact of alluvial upland with the meander-cut plain and the dry valley. Further, two hill outcrops of the Aravalli System also occur as upstanding residuals within the alluvial tract. These hills are narrow, flat-topped outliers rising to between 220 and 236 m above the mean sea level, and possess a relative relief of some 25 m. On the east-facing slope of one of these structural hills to the east of Firozepur distributary, a number of string-like sandy accumulations athwart the northwest-southeast crest directions have developed. These are generally less than one metre high deposits of fine sand, and probably developed as sand shadows in the zone of dead sand.

The nearly-level, southward sloping gentle surface of the Pleistocene upland tract is photo-interpreted to suffer from varying degrees of salt-hazards. Salt-afflorescences have been occasionally observed at the surface with the possible presence of calchie at shallow depth beneath the alluvial cover. The salt-effects appear to have been caused by poor surface drainage, and perpetuated due to application of brackish subsurface water for limited irrigation.

## Cover Types

The spatial pattern of cover-types in Firozepur Jhirka region is shown in Fig. 5.2, and its areal extent analysed in relation to the three geomorphic environments (Table 5.2). Thus, this study of land cover data is based on individual land resource units that comprise the three geomorphic environments.

## Inland Lake Evaporite Environment

Salt-affected, and waterlogged and potentially

waterlogged lands are the most typical and widespread cover-types in the 'lake flat' environmental resource unit. These respectively account for 83.63 and 10.94 per cent of the area analysed in the inland lake evaporite environment. The genesis of salt-affected lands was previously discussed in detail for the inland lake evaporite environment of the Nuh region. Though data on soils, excepting a fraction of the area that lies within the Gurgaon Canal Command, are not available for the entire region, it is by analogy suggested that these salt-affected lands also comprise of very deep, calcareous and non-calcareous, fine to coarse loamy soils of typic ustochrepts taxonomic group. A part of this land in Gurgaon Canal Command area has been classified by CAD (April 1986) in Class III land capability, where poor surface drainage, slight to moderate salinity and/or alkalinity hazards, low organic carbon and fertility status, and difficulty of tillage are inherent limitations of this type. The fertility status of these soils can be increased by improving drainage from flat lands and by avoiding over-irrigation; by using farm yard manure and green manure for the improvement of soil structure; and by the judicious application of gypsum for reclaiming salt-affected lands.

The waterlogged and potentially waterlogged lands confine along the drain passing near the Firozepur Jhirka town, and shallow depressions around the structural hills and within the lake flat. The soils are referred to as generally clayey, calcareous and non-calcareous, mixed hyperthermic vertic camborthids and aquic vertic camborthids. These areas are also subject to seasonal stagnation of water during the monsoon season.

TABLE 5.2: Area in Different Cover Types, Including Built-up Land in Firozepur Jhirka Region

| Geomorphic environment | Resource unit | Cover type | Area (sq km) |
|---|---|---|---|
| Inland Lake Evaporite Environment | Lake flat | Salt-affected land | 40.95 |
| | | Waterlogged and potentially water-logged land | 5.36 |
| | | Orchard/grove | 1.19 |
| | | Brick kiln | 0.35 |
| | | Wasteland, including village common land | 0.27 |
| | Structural hill | Aravalli hills | 0.84 |
| Dryland Environment | Alluvium | Alluvium | 1.01 |
| | Sandy plain | Sandy plain | 41.32 |
| | | Waterlogged and potentially water-logged land | 0.28 |
| | | Rocky protuberance | 0.32 |
| | Fixed dune | Fixed (stable) sand dune | 0.83 |
| | Structural hill | Aravalli hills | 45.29 |
| | | Piedmont | 41.62 |
| | | Broken land | 2.57 |
| | | Orchard/grove | 0.41 |
| | | Wasteland, including village common land | 9.98 |
| Fluvial Environment | Meander-cut plain | Undulating plain | 27.84 |
| | | Waterlogged and potentially water-logged land | 2.55 |
| | | Wasteland, including village common land | 0.83 |
| | | Grazing land | 0.46 |
| | Dry valley | Shallow valley | 16.90 |
| | | Orchard/grove | 0.06 |
| | Upland | Salt-affected land | 180.03 |
| | | Waterlogged and potentially water-logged land | 58.05 |
| | | Wasteland, including village common land | 4.80 |
| | | Marsh | 0.63 |
| | | Fixed (stable) sand dune | 0.48 |
| | | Rocky protuberance | 0.32 |
| | | Grazing land | 0.32 |
| | | Brick kiln | 0.10 |
| | | Mound | 0.01 |
| | Structural hill | Aravalli hills | 1.07 |
| | | Piedmont | 9.02 |

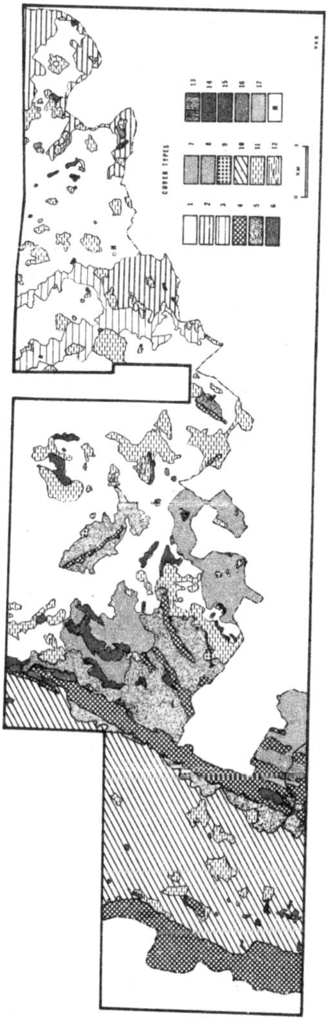

FIG. 5.2: Cover types in Firozepur Jhirka region. 1. Pleistocene alluvial upland, 2. Meander-cut plain, 3. Dry valley, 4. Aravalli hills, 5. Piedmont, 6. Broken land, 7. Alluvium, 8. Sandy plain, 9. Fixed sand dune, 10. Lake flat, 11. Waterlogged and potentially waterlogged land, 12. Marsh, 13. Rocky protuberance, 14. Wasteland, 15. Brick kiln, 16. Orchard/grove, 17. Grazing land, M. Mound.

The development of resource potential of the lake flat environmental 'unit, especially for agriculture, is largely contingent up on adequate provision of minor irrigation. Most of this unit has a marginal quality groundwater where the EC varies between 2000 and 6000 mmhos/cm. Pockets and areas of shallow fresh and shallow saline groundwater, and areas of highly sodic subsurface water have also been reported in parts of Firozepur Jhirka development block (CAD, April 1986). The application of these saline and alkali waters has further aggravated salt-hazards in the land that is inherently salt-affected.

The structural hill outliers are surrounded by an enormous thickness of the lake sediments. These rugged hills with skeletal, undifferentiated soils are of no consequence to agricultural use. Further, their small-size also inhibits any other economic activity. Resources of these barren hills may, however, be conserved by adopting suitable land management practices, such as afforestation.

## Dryland Environment

The dryland environment represents a geomorphic landscape dominated largely by structural hills, and to lesser extent by sandy accumulations. The alluvial cover is but an insignificant aspect of this environment.

The sandy plain resource unit comprises of undulating sandy plain, waterlogged and potentially waterlogged land, and rocky protuberances as cover types. Of these, the undulating sandy plain with 98.56 per cent of the area is by far the most prominent land cover in this environmental unit. It has very deep, well-drained sandy loam to loamy sand soils of low moisture holding capacity. The soils are also low in organic carbon and fertility. Further, application of irrigation is also difficult on the undulating surface of this cover type. Crop failures are common due mainly to uncertain rain and shifting sands. In terms of soil taxonomic classification, the soils of undulating sandy plain may be referred to as severely wind eroded, mixed, hyperthermic, typic ustipsamments. Soils fertility of this cover type may be enhanced by using appropriate quantities of farm yard manures, and by devising methods whereby soil moisture can be conserved for sufficient period of time. Wind erosion can be checked by providing wind breaks, and planting of perennial grasses within the undulating sandy plain cover type. The presence of potentially waterlogged tract within the sandy plain may perhaps indicate impervious strata at shallow depth from the surface.

The fixed or stable dunes are typical or excessively drained, non-calcareous hyperthermic family of moderately eroded ustipsamments. These hummocky dunes of fine sandy texture are sensitive to wind deflation and, therefore, should be prohibited for grazing activity or cultivation practices.

Aravalli hills, piedmont, broken land along stream courses debouching through the hills, orchard/grove, and wasteland have been identified as the five major cover types within the 'structural hill' environmental resource unit that encloses the Firozepur Jhirka Valley. Having witnessed episodic erosion over a long geological time span, the rugged Aravalli hills spread over 45.34 per cent of this environment. These denuded hills of moderate to steep slopes, and with skeletal soils, are not conducive for agricultural pursuits. Further, disintegrated and decomposed debris from hill slopes are produced in abundance. These products of weathering are subsequently made available for wind and water transport. It is, therefore, desirable that suitable land and water management practices are adopted to check the menace of denudation. In so doing, not only is the ecological environment recuperated, but also the natural potential of these lands for tourism can be exploited.

Piedmont is the second largest cover type with 41.67 per cent of the area in 'structural hill' resource unit. It forms a detritus covered gentle to moderately sloping surface at the foot of hill remnants and the eastern hill range. The shallow surficial deposits are coarse textured with ubiquitous distribution of rock fragments overlying a pale-brown regolith. The piedmont is a low productivity status unit of landscape due primarily to the nature of soils, and limitations of slope and relief.

Of no economic value are the broken lands that traverse the Aravalli slopes. This cover type represents a case of extreme environmental degradation. Hence, watershed approach is required for conservation of the hill slopes.

Intense denudation of the 'structural hills' has formed almost level surfaces of hard rock stratum dispersed at the foot of hills, within piedmont, and sandy plain environmental resource units. These surfaces, referred to as wasteland cover type, are devoid of a suitable environment for crop cultivation.

### Fluvial Environment

The fluvial environment represents a higher degree

of heterogeneity in topography resulting from the Recent activity of shifting course of the Jamuna river to inland drainage, and from sandy accumulations to structural hills. These provide large enough variations in the potentiality and productivity of land covers and, consequently, in the development of land resources of this region. In this section, the more prominent land covers of individual resource units have been summarised.

Nearly 96 per cent of the area in 'meander-cut plain' resource unit is occupied by an undulating plain (87.88), and waterlogged and potentially waterlogged land covers (8.05). The undulating plain is one of irregular slope and micro-relief variations resulting from lateral migration of the Jamuna river through short-segmented meander loops. This land cover provides excellent surface drainage, is free of salt hazards, and is characterised by fresh-quality groundwater at shallow to medium depths. Here, the soils are loamy sand in texture, and very deep. These soils, as reported in Chapter 2, overlie brown to yellowish coloured sands of aeolian origin, are referred to as Hiranki soil series (CAD, April 1986). Consequently, the undulating plain possesses moderately high primary productivity, suitable for supporting diversified and irrigated crops. The waterlogged lands are generally confined to the unit's boundary with the upland resource unit. These areas comprise of imperfectly drained sodic clayey loams developed in broad depressional areas in the alluvium. The soils may be analogous to be Khampur series of the CAD (April 1986).

Covering an area of 16.90 sq km, the shallow valley is by far the single largest geobiological unit of the dry valley resource unit. Here, excessively drained, very deep loamy soils of fluvial environment have developed on 1-5 per cent slope. These soils normally contain adequate soil moisture throughout the year from subsurface flow into this land cover from the adjacent irrigated tracts on Pleistocene upland surface, and from linear and lateral overland flow due to rainfall. Further, a salt-free soil environment has also been photo-interpreted. Coupled with the availability of fresh-quality groundwater, the shallow valley land cover is ideally suited for raising a number of high yielding varieties of crops under irrigated and rain-fed conditions.

Of the nine cover-types photo-interpreted in the Pleistocene alluvium, the salt-affected (73.56%), and waterlogged and potentially waterlogged (23.72%) lands are the most widespread. Salt-effects are observed to be most conspicuous along canal irrigation network and drains that pass through this environmental resource unit. Complete information on the soils of this region is not available. However, a western section of the resource unit falling within the Gurgaon Canal Command area is reportedly in coarse to fine loamy, calcareous and non-calcareous, mixed hyperthermic typic ustochrept family of soils (Goyal *et al.*, 1978). Some other soil series described for this area by the above referred authors could not, however, be recognised in this research. These soils are low in organic carbon, and possess low to moderate fertility status. Appropriate use of farm-yard manure and fertilizers, soil moisture conservation, and gypsum treatment of the soils can further the fertility level of the salt-affected lands. Here, the subsurface water is marginal-saline, but can be suitably used in conjunction with the canal water for minimising the salt hazards and for raising crops of low to medium salt tolerance.

Waterlogged and potentially waterlogged lands

have prominently developed along the south-directed surface drains and a few canals, and within local depressions in the alluvium. The last named usually become inundated between July and September months of the summer monsoon season. It is in some of the very severally waterlogged covers that marshes, covering 0.28 sq km area, have established. Given a semiarid climatic regime of the area, this land cover would indicate high water salinity and associated salt hazards around its margins. In fact, the quality of groundwater around the waterlogged and potentially waterlogged lands is shown to be saline, which in many cases is indicated to be unfit for drinking. The areas subject to seasonal inundation can, however, support *rabi* season crops of low salt tolerance.

The structural hill resource unit comprises of the Aravallis and piedmont cover types with their specific geobiological environments. Their potential and capability have been previously described in the context of 'inland lake evaporite' and 'dryland' environments.

## Wasteland in the Firozepur Jhirka Region

The wasteland, classified into culturable and non-culturable types, covers 23.93 per cent of the total area of the Firozepur Jhirka region (Table 5.3). Of this, 17.54 per cent has been recognised in culturable and the remaining 85.21 per cent in non-culturable wasteland category. Most of the area in culturable wasteland is due to neglect/excessive salts, and can be reclaimed by using techniques specific to the needs of the three geomorphic environments. Similarly, wasteland due to abandoned brick kilns may also be reclaimed for subsequent use. The environmental considerations do not, however, allow reclaiming of a total of 3.14 sq km in wasteland due to limitations of slope, wet-lands and stable dunes. The non-culturable wasteland is the highest in the dryland environment and alone accounts for 86.87 per cent of the total non-culturable wasteland of the Firozepur Jhirka region. These lands are in the form of hills, piedmont, rocky protuberances, and/or broken land which impose severe restrictions on expansion and development of agriculture and related infrastructure in the region.

## Agricultural Land-use

As abiotic elements of landscape, the geomorphic environment has its own inherent potential and limitation for supporting specific crops at a certain optimum productivity level. In the present context, the dynamic interaction between abiotic and biotic parameters of relevance to crop culture has been examined for randomly selected four villages in the three geomorphic environments discussed in this chapter. The yield potential and productivity of crops is the result of utilisation and management of abiotic factors of the geomorphic environment, and the technology developed to exploit the natural potential of the geobiological environment.

## Inland Lake Evaporite Environment

Bhaikhera (S.O.I. topog aphical map $54^E/1$ has been randomly selected as a village representing the geobiological complex within the 'lake flat' land cover. Its geomorphic characteristics, resource potential and limitations were briefly discussed in the previous section.

TABLE 5.3: Classification of Wasteland in Ferozepur Jhirka Region

*[Area (sq km)]*

| Geomorphic environment | Type/ class* | Wasteland due to | | | | | | | | | |
|---|---|---|---|---|---|---|---|---|---|---|---|
| | | Neglect/ excessive salts | Slope | Swamp | Fixed dune | Abandoned brick kiln | Hill | Pedi- | Rocky protuberance | Broken land | Total |
| Inland lake evaporite environment | 1 | 0.27 | — | — | — | — | — | — | — | — | 0.27 |
| | 2 | — | — | — | — | — | 0.84 | — | — | — | 0.84 |
| Dryland environment | 1 | 9.61 | — | — | 0.83 | 0.27 | — | — | — | — | 10.71 |
| | 2 | — | — | — | — | — | 45.29 | 41.62 | 0.32 | 2.57 | 89.80 |
| Fluvial environment | 1 | 3.95 | 1.20 | 0.63 | 0.48 | 0.30 | — | — | — | — | 6.56 |
| | 2 | — | — | — | — | — | 1.07 | 9.02 | 0.32 | — | 10.41 |
| Total | 1 | 13.83 | 1.20 | 0.63 | 1.31 | 0.57 | | | | | 17.54 |
| | 2 | — | — | — | — | — | 47.20 | 50.64 | 0.64 | 2.57 | 101.05 |

* 1. Culturable wasteland; 2. Non-culturable wasteland.

The 1976-77 inventory of crops in the village shows the dominance of *kharif bajra* and *Jowar* as fodder crops. The *rabi* season crops in the decreasing order of area importance are wheat, barley and *sarson*. Irrigation facilities are poor, inhibiting the production potential of crops per unit area per unit of time. Irrigation, all of which is based on exploitation of marginal quality groundwater, was only 12 per cent of the net cultivated area and resulted in a cropping intensity of 120 per cent in the year 1976-77.

*Bajra* is a high-yield low-spread crop of the region. A few hybrid varieties of *bajra* that have low water requirement, and can be suitably grown under conditions of moderate to heavy soils, may be considered for this resource unit. Hybrid *bajra* is also adaptable for rotation with dwarf wheat under irrigated conditions. The yield of *bajra* per unit area can, therefore, be increased manyfold provided improved and adequate irrigation and input facilities are introduced in this region.

*Jowar* and *sarson* are, in particular, salt-sensitive crops, but in the heavier soils of this region they can withstand a higher alkalinity than is possible in the coarse-grained soils. Here, *jowar* is a low-yield and low-spread crop, and *sarson* is a low-yield but high-spread crop. Both these low-yield crops are, however, grown in the 'lake flat' cover type due to their low water requirement. On the other hand, barley is a high-yield and high spread crop as it is ideally adapted to the high salt environment of the soil.

The yield potential of the 'lake flat' can be enhanced by providing increased irrigation facilities possibly through conjunctive use of surface and subsurface water resources, by treating soils with gypsum, calcium chloride, and sulphur, by adding green and farm-yard manures, and by deep ploughing and trenching to improve subsurface drainage, and by the intelligent management of soils. With these practices, the productivity level of the resource unit can be enhanced by adopting a multiple cropping system of *bajra*-cowpea-*guar* (fodder) in the summer, and pulses-wheat-barley-raya in the winter season. This environment is also extremely suitable for raising paddy, but its cultivation will have to wait till such time that adequate irrigation delta can be created for the support of this crop.

**Dryland Environment**

Manota (S.O.I. topographical map 54 $^{E}/1$ is situated at the boundary of dryland and fluvial environments, but the cultivated area of the village is largely confined to be undulating sandy plain cover type. In this village, *bajra* and *jowar* as fodder are the principal *kharif* crops, while wheat, *sarson* and fodder, in this order of importance, constitute the crops of the *rabi* season. The intensity of irrigation is determined to be 10 per cent, in which the minor irrigation is the only mode of irrigation. For the same year of 1976-77, the intensity of cropping was computed at 129 per cent.

The above crops are in close conformity with the geobiological environment of the resource unit and the undulating sandy plain land cover. The use of marginal quality groundwater for irrigation, poor moisture retention capacity of soils, and tillage of sandy soils leading to a greater risk of wind deflation are, however, detrimental to realising the optimum potential of the cover type for increasing the productivity of the above crops. Controlled sprinkler irrigation is perhaps the only good solution for minimising the above said ill-effects, and for raising the productivity level of the land resource unit. For example, hybrid bajra under irrigated conditions can easily replace the local varieties which despite a large areal extent, have rather lower yield per unit area. Similarly, the output of both wheat and *sarson* in the salt-free environment of the undulating sandy plain can be increased by adopting suitable methods of soil moisture conservation.

### Fluvial Environment

From among the four major land covers in the fluvial environment, Nimka and Raipur were selected in the Pleistocene upland cover type to study the agricultural use of land in this geobiological complex. Nimka and Raipur are respectively situated to the north of 'shallow valley' land cover (S.O.I. topographical map $54^{E}/5$), and between the Gurgaon Canal and the Hattin-Bharatpur Escape (S.O.I. topographical map $54^{E}/1$). As both the villages are situated in a similar environment, they should theoretically present similar crops. The intra-regional variations, if any, may be due to differences in the infrastructural variables for the support of agricultural operations.

Nimka raises *bajra, jowar* and fodder in the *kharif* and wheat, sarson and barley in the *rabi* season. Irrigated agriculture is supported by minor irrigation which provided 38 per cent intensity of irrigation and resulted in 135 per cent intensity of cropping. Likewise, the prominent *kharif* crops at Raipur in 1976-77 were *bajra, jowar* and fodder; and wheat, *sarson* and barley provided for the crops of the *rabi* season. Irrigated agriculture in this village is supported by both canal and minor irrigations. The intensity of irrigation and the intensity of cropping for the year under reference are 54 and 166 per cent respectively.

Despite infrastructural differences for the support of agricultural activities, the nature of crops and the cropping pattern in the two village remains remarkably similar. Data on the yield of crops per unit area are, however, not available to suggest if the productivity of crops also differ due to the above-said intra-regional differences in the intensity of irrigation. Notwithstanding the similarity in the type of crops grown, the fact clearly highlights that the geomorphic environment in this region is the primary determinant of agricultural activity.

Given the agro-environment of the resource unit and the cover type, it seems the higher productivity can be realised only by the increasing dependence on crop planning under limited water supply conditions. The yield potential of *kharif* crops like *bajra* and *jowar* can be increased by inter-row water harvesting techniques. Sprinkler irrigation is also appropriate for maximising the yield of *rabi* crops. The crops grown in these two villages are low to medium water requirement crops, and in the semiarid climatic environment where long periods of drought commonly occur, timely application of available irrigation water, though limited, can save crops from total failure. Besides, due to higher grain and straw yields *bajra* would continue to be the major crop of this region.

# 6

# SUMMARY AND CONCLUSIONS

APPRAISAL OF land resources is an essential precondition for a long-term rational land-use planning. For rural-based economies, such as in India, land-use planning must basically address itself to the planning of agricultural land-use so that the inherent potential of land resources can be fully developed, exploited, and conserved for sustained economic growth rate necessary for improving the quality of life of some 80 per cent of the population dependent directly or indirectly on agricultural pursuits. It is in this context that the land resources in some 5483 sq km area dispersed in parts of Sonipat, Rohtak, Bhiwani, Mahendragarh, and Gurgaon (including Faridabad) districts of southern Haryana and a fraction of territory in the adjoining Delhi have been classified and evaluated for the inherent potential and limitation *vis-a-vis* agricultural land-use. These areas support a variety of crops under different agro-environmental conditions and under varying infrastructural facilities. Consequently, the yield of crops per unit area and economic returns to farmers from the cropped areas vary spatially in relation to the productivity potential of the land resources.

Until a few years ago, property boundaries of land resources were demarcated and land capability determined by time-consuming field surveys. In the recent years, remote sensing applications have provided the inventory data expeditiously and economically direct from interpretation of aerial photographs, satellite imagery, and airborne radar observations. These techniques are now widely used for mapping and preparation of accurate land inventory maps with limited field-checks. Aerial photographs in particular have been extensively employed for land resource evaluation, for mapping land capability classes, and for delimiting soil-type boundaries.

In the present study, black and white panchromatic prints on an approximate scale of 1: 25,000 have been exclusively used for delimiting land resource property boundaries. Land resources are analysed on the basis of standard photo elements, such as photo texture (smoothness or coarseness of images), tonal contrasts, relative shape and size, associations, and convergence of evidence. The boundaries of land resource units were field-checked along selected traverses, and transferred on to the respective Survey of India topographical maps on 1: 50,000 scale.

A number of land classifications for a variety of purposes have been proposed in literature. Some such terrain and soil-based classifications have been briefly discussed, and the one based on geomorphic criterion was determined to be the most appropriate for application to the evaluation of land resources in the study region. In the present case, geomorphic processes that are operating presently or have operated in the past, the morphologic attributes and associated sedimentary properties, and subsurface geological properties of sediments have been considered as diagnostic parameters of land capability and potentiality. Thus, three dominant geological environment – 'fluvial', 'dryland' and 'inland lake evaporite' – have been recognised in parts of southern Haryana and Delhi. The above-said geological environments have been further subsumed into genetic ensemble of landforms, each of which are a product of a dominant geomorphic process, and are also homogeneous with respect to morphology and material composition. These landforms have been referred to as 'environmental resource units' or geobiological resource units' as they are also associated with specific hydrologic and vegetation characteristics. Further, these resource units have been observed to possess uniform inherent productivity potentials, and respond in almost a like-manner to the application of infrastructural facilities for enhancing the productive potential of environmental resource units.

Agricultural land-use planning integrates data on environmental resource units, land covers or cover types, and agricultural land-use. In the study area, where the physical limits of agricultural land-use have been practically reached, the photo-interpreted land cover data included diverse information on the association of environmental resource units with other characteristics of land. These included salt-affected land, waterlogged and potentially waterlogged land, marsh land, water bodies, and Class-II data on other uses of land, such as brick-making activity, orchard/grove, grazing land, and various classes of wasteland, to name a few.

Cropping pattern of randomly sampled villages from the selected environmental resource units was analysed for the year of photography, and related with the geobiological complex within which the village is situated. The purpose of this study was to interpret the effect of environment on crops and crop production, and to suggest ecologically stable and economically productive crop-land use of environmental resource units.

No attempt has been made in the present research to delimit the major soil-types and land capability classes. The reports available for parts of the study area, however, suggests that these two very closely describe the attributes of environmental resource units defined and discussed in this research. The classification of land resources on the bases of geomorphic criterion can, therefore, also be developed further to study the two above-said components of land resource evaluation.

Region-wise selected physico-chemical attributes of environmental resource units are summarised below. The cropping pattern has been examined in relation to the geobiological environment and changes, if any, required in cropping pattern for maximising the productive use of environmental resource units are suggested. Wasteland have also been classified and quantitatively analysed.

The Sonipat region is an area of fluvial landscape divisible into three Quaternary sub-periods and four environmental resource units. The photographic interpretation of the geomorphic landscape suggests that nearly all of the area studied is in reworked or 'newer' alluvium, and that little of the original Pleistocene upland tract can be recognised in this region. Low-relief longitudinal bars interspersed with broad shallow depressions have been identified as Sub-recent over flow channels adjacent to the Jamuna river. Composed of medium to coarse-grained bedload materials, this environmental resource unit possesses deep sub-surface drainage, is free of salt hazards and suitable for low productivity dryland farming of coarse foodgrains and creeping crops. Meander-cut plain and point bar accretions, the other environmental resource units, are of the Recent geological periods. The meander-cut plain is interpreted to have formed from major shifts in the course of river channels and through minor intersecting maze of meander loops. Brick-making industry has especially flourished along and within abandoned channel courses where generally clayey sediments are available for use in brick-making. Abandoned brick kiln areas have, however, added largely to the culturable wastelands of the resource units. This environmental resource unit is noted for repeated and irregular micro-relief variations, poor to good surface and subsurface drainage, fresh to marginal quality groundwater, salt-free and salt-affected lands and, waterlogged and potentially waterlogged conditions. The 1969-70 cropping pattern of salt-free lands that occupy well-drained higher topographic surfaces was found to be in conformity with the salt-free environment. The yield per unit area of crops like sugarcane and gram can, however, be enhanced by strengthening irrigation facilities. Oilseed, groundnut and citrus fruit crops can also flourish in this environment. The salt-affected land cover is a widespread occurrence within the meander-cut resource unit. The soils are reportedly moderately to highly saline throughout the depth of soil profile. The lands of this type occur in association with topographic lows and canal network channels, and support a relatively high proportion of culturable wastelands. The 1969-70 cropping pattern of *Kharif* coarse foodgrains and *rabi* wheat, grain, and *jowar* does not appear to be ideal for such an environment. It is perhaps with the intensification of soil-salts that the cropping pattern in 1986-87 evolved to rice-wheat-fodder-sugarcane-pulses. Higher productivity from this cover type can be achieved if a more suitable cropping pattern comprising of *dhaincha*, barley, *barseem*, paddy, wheat and cotton is more closely examined. Gypsum treatment of land is, however, essential for reclaiming the land for higher productivity status. Waterlogged and potentially waterlogged land cover represents a site-specific dynamic and unstable environment around irrigation channels, topographic depressions and poorly drained surfaces. In this case also, the *kharif* and *rabi* land occupancy respectively in fodder, *bajra* and sugarcane, and wheat and gram does not adequately relate with the high watertable of marginal to saline quality conditions. The crops, such as paddy, sugarcane and fodder that can withstand salt-hazards and also have a higher water demand can be more remunerative in this environment. The productive use of point-bar environmental resource unit comprising of sandy soils with large and rapid relief variations due to swales and depressions is rather limited. The monotonously flat expressionless surface of deep Pleistocene alluvium constitutes the upland-tract. This environmental resource unit in Sonipat region is one of insignificant extent. The productive potential of the upland surface is, however, determined primarily by the disposition and proportion of calchie in the soil profile.

The Rewari region is the domain of dryland environment represented by the Quarternary blown sand and alluvium, and the Precambrian structural hill outcrops. Undulating sandy plain, dune ridge complex, and fixed dune have been recognised as the three principal environmental resource units of the Quaternary blown sand. The undulating sandy plain of diverse relief composition represents neutral and sodic soils, belonging to hyperthermic, torripsamment and ustochrepts taxonomic groups. Soil sodicity is a recent phenomenon that can be traced to the developments in surface irrigation, and to the ever-increasing dependence of agriculture on exploitation of marginal quality groundwater for irrigation. Most significantly, these effects and waterlogged conditions have become equally more pronounced in *tals* or broad shallow depressions within the resource unit. This cover type is most intensively cultivated and irrigated within the undulating sandy plain resource unit. Here, the grazing lands are of marginal quality which require controlled grazing practices for checking the menace of deflation. All the land within the undulating sandy plain is reported in Class-III and Class-V types for which specific land and water management techniques are required for maintaining the ecological stability of the resource unit. This environmental resource unit supports coarse foodgrain crops under rainfed conditions in the *kharif*, and gram, oilseeds, and barley with limited irrigation in the *rabi* season. The productivity potential of the undulating sandy plain may be enhanced if rainfed *bajra*-gram crop sequence is replaced by *bajra*-oilseeds under similar conditions. In areas of limited irrigation, however, wheat-barley and wheat-*bajra* crop sequences are likely to give higher yields per unit area. The dune ridge complex is a very low productivity environmental resource unit that comprises of dune ridge, *tal*, rocky protuberance, alluvium of the Krishnawati stream, depression, waterlogged and potentially waterlogged land and wasteland as land covers, all of which area is reportedly in Class-IV land. Periodic crop failures on large scale, extremely limited irrigation and deflation by wind are the major constraints in land and water management requirements of this resource unit. It is suggested that the productive efficiency of crops can be slightly enhanced if priorities in limited irrigation are shifted from barley to gram. Rainfed *bajra* will, however, continue to be the dominant crop of this region. The existence of fixed dune environmental resource unit is essential to the preservation of ecological balance prevailing in the dryland environment of the Rewari region. It is, therefore, suggested that the morphological attributes of this geobiological unit spread over 48.90 sq km area are not at all disturbed. The environmental resource unit recognised as the Quaternary Sahibi alluvium is represented by eight cover types in ustochrepts, ustorthents, and natraqualfs varying in texture from sand, sandy silt to clayey loam. These soils are low in fertility status. Further, they also suffer variously from seasonal flooding, high watertable conditions, presence of soil salts and wet lands. The Sahibi alluvium supports a variety of crops related essentially to the physico-chemical properties of major land covers, and irrigation facilities from subsurface water resources of fresh-water quality. Rainfed *kharif* crops are grown on residual soil moisture, while the *rabi* crops are irrigated to some extent. It is suggested that judicious application of irrigation water is necessary for the furtherance of agricultural productivity of the resource unit. Thus, the overall performance of agriculture may be enhanced by adopting a crop sequence of rainfed *bajra*-irrigated wheat-rainfed gram-rainfed barley and irrigated oilseed crops that may be individually suited to the salt-free, salt-affected and potentially waterlogged soil environments. Development of excessive concentration of soil salts, wet lands and abandoned brick kilns have contributed to the culturable, and rocky protuberances to the non-culturable wastelands. Most of the structural hill environmental unit represents non-culturable wasteland in hill, pediment, and broken land categories. The culturable wasteland is given by widespread waterlogged conditions. Thus, less than 6 out of 220.44 sq km area in this resource unit is actually available for cultivation. The geomorphic landscape of exceptional beauty in the structural hill resource unit does, however, offer immense opportunity for further development of tourism. Mining activity is likely to stay as a small-scale local economic enterprise.

The dryland environment of Nuh region has been subsumed into two major geological periods and three physiographic provinces. Aeolian and still-water deposits of the Recent and Aravalli structural hills of Precambrian origin have been recognised. The Precambrian element of geomorphic landscape occurs as independent continuous ridge and also outliers within the Recent deposits. The aeolian environment comprises of a limited extent of alluvium deposited by larger rivulets draining the Aravalli hill slopes, extensive undulating sand cover overlying rockbed at shallow depth, broken land along rivulets, hill outliers, and Aravalli hill range. The extensively cultivated undulating sandy plain has sand to sandy loam textured ustipsamment soils. Here, salt-affected lands occur in association with high watertable conditions, and seasonal wetting and drying of surficial materials along stream courses and foothills with gentle slopes.

Barrages across streams passing through the sandy plain have created temporary water bodies, potentially waterlogged conditions, and soil salts. About 60 per cent of the surveyed area in this environment is in non-culturable wasteland category, and may be explained by the structural hills. It is, however, interesting to note that the culturable wasteland here is non-existent. The properties, potentialities and inherent limitations of the undulating sandy plain environmental resource unit for resource development are similar to the one described for the unit in Rewari region. The still-water deposits in the Nuh region represent a part of an extinct inland lake. Evaporite deposits and irrigation application of brackish subsurface water resources explain salt effects in nearly 82 per cent of the area. Another 13 per cent of the area in this region is in water bodies, marsh land and waterlogged and potentially waterlogged conditions. The culturable wasteland is made up of salt-affected land, and land in marsh, water bodies and abandoned brick kilns. The Aravalli hill outliers within provide for the non-culturable wasteland. The soils of the lake evaporite environment are generally loamy to clayey typic ustorthents and typic camborthids suffering from a variable degree of salt hazards. Available information for a part of this region classifies the land resources into three land capability types. Class II land is affected by slight alkali hazards and raises rainfed *bajra, jowar*, barley and *dhaincha*, while wheat, sugarcane and vegetables are grown under irrigated conditions. Class III and suffers from a host of inherent limitations, such as moderate salinity/alkalinity hazards, occasional stagnation of water during rainy season, medium to coarse-grained soils, and low moisture holding capacity of the soils. Here, the *kharif* (rainfed *jowar, bajra,* barley) and *rabi* (wheat, gram, and wheat-gram mixture under irrigated conditions) crops appear to be loosely adjusted with the soil environment. Salt-tolerant paddy, barley and sugarcane can replace the coarse foodgrain *kharif* crops with canal irrigation and, thus, enhance the productivity level of the resource unit. Adequate subsurface drainage is, however, necessary to prevent waterlogging, and intensification of soil hazards. Class VIII land spreads over eroded Aravalli hills, piedmont and rocky protuberances. In sum, suitable land, water, and crop management are required for enhancing the productivity potential of this environmental resource unit.

The Firozepur Jhirka region is an ensemble of all the three previously described geomorphic environments. Therefore, the physical attributes, inherent potential, and set of limitations in the development of land resources in each of the three geological environments of this region are shown normally similar to the given environment(s) in the other regions discussed in this research. Differences in the productivity levels of land resource do occur however, and are explained by the development of infrastructural facilities for the support of agriculture. Suitable cropping patterns and agronomic practices are also suggested for enhancing the productivity potential of environmental resource units in the Firozepur Jhirka region.

This study has shown that the spatial pattern of crops and cropping pattern is closely determined by the physico-chemical properties of the environ-mental resource units. The effect of technology-aided crops on environment has also been briefly examined as part of the explanation of man-environment interaction.

# BIBLIOGRAPHY


Anderson, J.R. and Whitmer, R.E. (1984), Land resource planning applications of land-use mapping and inventory from remotely- sensed data. *Eighteenth Internat. Symp. Remote Sensing of Environment*, Paris, Oct. 1-5, pp. 429-443.

Awasthi, S.C., M. Prasad and A. Chatterjee (1976), Extension of a lake and presence of evaporites in Nuh area, Gurgaon district, Haryana. In *Quaternary: Recent Geological Environments and Environmental Management*, Vol. III Geol. Surv. India (mimeo).

Baweja, B.K. and Sharma S.K. (1978), Role of remote sensing in groundwater surveys and exploration in India. *Proc. Joint Indo-US Workshop on Remote Sensing of Water Resources*, Hyderabad, National Remote Sensing Agency, April 10-14, pp. 68-77.

Beckett, P.H.T., R. Webster, G.M. McNeil and C.W. Mitchell (1972), Terrain evaluation by means of a data bank. *Geogr. J.*, 138, 430-449.

Bhargava, G.P. (1977), A report on soil salinity at the government agricultural farm Nayabans (Rohtak), Haryana. Karnal, Div. Soils and Agro., Central Soil Salinity Research Institute, Rept. No. 4, 15 pp. (mimeo).

Bhargava, G.P. (1985), Soil characterisation and salinity appraisal for horizontal drainage at

Sampla. Karnal, Div. Soils and Agro., Central Soil Salinity Research Institute, Rept. No. 10, 61 pp. (mimeo).

Bhargava, G.P. and Sharma. D.R. (1984), Soil characterisation and salinity appraisal at Mundlana. Karnal, Div. Soils and Agro., Central Soil Salinity Research Institute, Rept. No. 9, 60pp. (mimeo).

Bridges, E.M. and Doornkamp, J.C. (1963), Morphological mapping and the study of soil patterns. *Geogr.*, 48, 175-181.

Budowski, G. and Miller, K.R. (1973), North and Latin America: Comparisons and contrasts in evaluating land for nature conservation. In A.B. Costin and R.H. Groves (Eds.) *Nature Conservation in the Pacific*, Canberra, ANU Press, pp. 27-38.

Christian, C.S. (1958), The concept of land units and land systems. *Proc. Ninth Pacific Cong.*, 20, 74-81.

Christian, C.S. (1959), The eco-complex in its importance for agricultural assessment. In *Monographiae Biologicae*, VIII, 587-605.

Christian, C.S. and Stewart, G.A. (1952), Survey of Katherine-Darwin Region, 1946. C.S.I.R.O., *Land Res. Ser.* No. 1.

Cooke, R.U. (1977), Applied geomorphological studies in deserts: A review of examples. In J.R. Hails (Ed.) *Applied Geomorphology*. Amsterdam, Elsevier Sci. Publ. Co., pp. 183-225.

Cooke, R.U., A.S. Goudie, and J.C. Doornkamp (1978), Middle East Review and bibliography of geomorphological contributions. *Q. J. Engng. Geol.* 11, 9-18.

Dhankar, R.P., S.P. Jain, T. Lall, K.P.C. Rana, and S. Pandey (1983), *Geomorphology, Soils and Land*

*Use of Haryana.* Nat. Bureau of Soil Surv. and Land Use Planning, *Nagpur. Bull.* No. 9, 21 pp.

Dow, V., H. Kienholz, M. Palm, and J. Ives (1981), Mountain hazard mapping: The development of prototype combined hazards map, Monarch Lake Quadrangle, Colorado, U.S.A. *Mt. Res. Dev.*, 1, 55-64.

Geddes, A. (1960), The alluvial morphology of the Indo-Gangetic Plain: Its mapping and geographical significance. *Trans. Inst. Br. Geogr.*, 253-276.

Glennie, K.W. (1970), *Desert Sedimentary Environments.* Developments in Sedimentology, 14, Amsterdam, Elsevier Sci. Publ. Co., 222 pp.

Goudie, A.S., B. Allchin, and K.T.M. Hegde (1973), The former extension of the great Indian sand desert. *Geog. J.*, 243-257.

Government of Haryana, Command Area Development Authority (April 1986), *CAD Project Report for Gurgaon Canal Command.* Hissar, 218 pp. (mimeo).

Government of Haryana, Command Area Development Authority (October 1986), *CAD Project Report for J.L.N. Lift Irrigation Canal Command.* Hissar, 188pp. (mimeo).

Government of India, Department of Space, National Remote Sensing Agency (1985), *Mapping of Wastelands in India from Satellite Imagery* (1980-82- Summary Report): Haryana and Delhi, 8 pp.

Government of Punjab (1916), *Punjab District Gazetteers*, Vol. IIA: Hissar District (Part A) 1915. Lahore, Punjab.

Goyal, V.P. and Singh, M. (1982), *Soils of Rohtak District (Haryana)*, Hissar, Haryana Agricultural University, 64 pp.

Goyal, V.P., M.S. Kuhad, and A.K. Jain (1978), *Soil Survey of Gurgaon Canal Command Area, Gurgaon (Haryana).* Hissar, Haryana Agricultural University, Bull. No. 6, 96 pp. (mimeo).

Gregory, A.F. (1973), A possible Canadian role in future global remote sensing. *Can. Aeronautics and Space J.*, 19, 85-92.

Gupta, M.C. (1983a), Groundwater quality. In *Studies for the Use of Saline Water in the Command Areas of Irrigation Projects Haryana, India.* Haryana State Minor Irrigation (Tubewells) Corporation Ltd. Workshop 28 July 1983, pp. 42-50.

Gupta, M.C. (1983b), Groundwater Conditions. In *Studies for the Use of Saline Water in the Command Areas of Irrigation Projects Haryana, India.* Haryana State Minor Irrigation (Tubewells) Corporation Ltd. Workshop 28 July 1983, pp. 58-65.

Hardie, L.A. (1984), Evaporites: Marine or non-marine? *Am. J. Sci.*, 284, 193-240.

Millard, R.S. (1973), Discussion on terrain by means of a data bank. *Geogr. J.*, 138, 450-451.

Ollier, C.D. (1977), Terrain classification: Methods, applications and principles. In J.R. Hails (Ed.) *Applied Geomorphology.* Amsterdam, Elsevier Sci. Publ. Co. pp., 227-316.

Parmar, A.S. (1986), Flow Patterns in Sahibi Basin, Haryana. Kurukshetra University M.Phil. Dissertation (Unpubl.).

Pascoe, E.H. (1950), *Manual of the Geology of India and Burma*, V.3, 3rd Edn. Govt. of India Press, Calcutta.

Raghavaswamy, V. (1982), Role of satellite remote sensing in land system mapping: Land resource inventory and land-use planning – A sample study of Kemang river basin, Arunachal Pradesh. *Photonirvachak*, 10(3).

Raghavswamy, V. and Vaidyanadhan, R. (1980), Morphology and land systems of a part of Visakhapatnam district, Andhra Pradesh from air photo studies. *Photonirvachak*, 8, 11-21.

Ramaswamy, M., S.M. Sinha, and P.C. Dev (1984), Environmental evaluation of Madchal taluk, A.P. using aerial photographs. *Photonirvachak*, 12, 39-46.

Rani, R. (1978), Rainfall Patterns of Haryana. Kurukshetra University M.Phil. Dissertation (Unpubl.).

Rao, M.B.R. (1973), The subsurface geology of the Indo-Gangetic Plain. *J. Geol. Soc. Ind.*, 14, 217-242.

Rawat, B.S., V. Rengamannar, and O.P. Joshi (1988), The appraisal of geo-environment in parts of Zahirabad area, Andhra Pradesh, using remote sensing techniques and collateral data. *Proc. Nat. Symp. Remote Sensing in Rural Development*, Nov. 17-19. Hisar, H.A.U., pp. 83-92.

Sastri, V.V., L.L. Bhandari, A.T.R. Raju, and A.K. Dutta (1971), Tectonic framework and subsurface stratigraphy of the Ganga Basin. *J. Goel. Soc. India*, 12, 222-233.

Shankaranarayan, K.A. (1985), Monitoring of desertification: A conceptual frame. In A.K. Tewari (Ed.) *Recent Advances in Monitoring the Process of Desertification*, Vol. 1. Nat. Workshop, Univ. Jodhpur Oct. 27-Nov. 3, 1985, pp. 9-26.

Sharma, R.L. '(1980), Dry Farming in the Arid Zone of Haryana - A geographical Perspective. Kurukshetra University Ph.D. Dissertation (Unpubl.).

Sharma, V.K. (1986), Geomorphic criteria in land classification. In A.B. Mukerji and A. Ahmad (Eds.) *India - Culture, Society and Economy*, New Delhi, Inter-India Publications, pp. 601-608.

Sharma, V.K. (1986a), *Geomorphology*, New Delhi, Tata Mcgraw-Hill, 244 pp.

Sharma, V.K. (1988), Environmental geology of a part of the Indo-Gangetic Plain in Haryana and Delhi. *Nat. Conf. Env. Degradation and Regeneration*, Patiala, Punjabi University, Oct. 28-30, 1988, 40 pp. (mimeo).

Srivastava, A.K., P.M. Jalota, G. Ghosh, S. Anantharaman, and A.K. Singh (1976), A review of geotechnical studies and its application in environmental management of Delhi. In *Quaternary: Recent Geological Environments and Environmental Management*, V. III, Goel. Surv. India (mimeo).

Sinha, R.K. (1986), Morphotectonic evolution of the Quaternary landsystems of the middle reaches of Yamuna Basin. *Intern. Symp. Neotectonics in South Asia*, Dehra Dun, Feb. 18-21, pp. 289-300.

Spangle, W. and Associates, F. Beach Leighton and Associates, and Baxter, McDonald and Company (1976), Earth-science information in land-use planning - Guidelines for earth scientists and planners. *U.S. Geological Survey Circular* 721, 28 pp.

Troll, C. (1965), Seasonal climates of the earth. In E. Rodenwaldt and H. Jusatz (Eds.) *World Maps of Climatology*, Berlin, Springer-Verlag, 28 pp.

United Nations, Food and Agricultural Organisation (1974), *Approaches to Land Classification*, Bull. 22, 120 pp.

United Nations, Food and Agricultural Organisation (1976), A *Framework for Land Evaluation*. Bull. 32, 72 pp.

Van Lopik, J.R. and Kolb, C.R. (1959), A *technique for Preparing Desert Terrain Analogs*. Vikasburg, Miss., U.S. Army Crops Eng. Waterw. Exp. Sta. Tech. Rept., 506 pp.

Waters, R.S. (1958), Morphological mapping. *Geogr.*, 43, 11-17.

Way, D.S. (1973), *Terrain Analysis*. Stroudsburg, Pa., Dowden, Hutchinson and Ross, 392 pp.

Williams, M.A.J. (1985), Pleistocene aridity in tropical Africa, Australia and Asia. In I. Douglas and T. Spencer (Eds.) *Environmental Change and Tropical Geomorphology*. London, George Allen & Unwin, 219-233.

Wright, R.L. (1972), Principles in a geomorphological approach to land classification. *Z. Geomorph.*, 16, 367-373.

# INDEX